JCER

Journal of Consciousness Exploration & Research

Volume 11 Issue 5
August 2020

Observational Panpsychism, Phenomenon of Memory, Hypno-Channelings, & Reconciliation of Consciousness

Editors:

Huping Hu, Ph.D., J.D.

Maoxin Wu, M.D., Ph.D.

Advisory Board

ISSN: 2153-8212   Journal of Consciousness Exploration & Research   www.JCER.com
Published by QuantumDream, Inc.

(Published in Journal of Consciousness Exploration & Research | August 2020 | Volume 11 | Issue 5 | pp. 438-551)

Table of Contents

Articles

Research Essays

Explorations

Perspective

Dialogue

(Published in Journal of Consciousness Exploration & Research | August 2020 | Volume 11 | Issue 5 | pp. 438-551)

Table of Content

Book Review

Review of Wolfgang Baer's Book: Conscious Action Theory - An Introduction to the Event-Oriented World View
James H. Lake

Article

Consciousness As a Phenomenon of Memory

Henry Grynnsten[*]

Abstract

Consciousness has been a mystery for thousands of years. The idea behind this article is that consciousness can be explained by way of the phenomenon of déjà vu. This state seems to come about when a synchronization error appears in sense signals going to the brain. From déjà vu, where there seems to be a memory of the present moment, one can work out that consciousness is a function of memory. Data from the senses are separate, while we see the world as a total experience. This means that the data are bound together in some way. Human consciousness can be explained as a phenomenon that arises when sense data are repeated at a short interval in the brain. The reason that humans have consciousness may have to do with language learning. Consciousness possibly appeared simultaneously with language, in a short span of time, and it is consciousness that makes human language possible. There was such a great advantage to using language, that all Homo sapiens acquired it.

Keywords: Consciousness, memory, phenomenon, language, human.

1. Introduction

The question of consciousness has been described as a hard problem, a term famously coined by David Chalmers [1]. It often seems that the subjective experience of consciousness influences our theories about it, so that it becomes hard to put your finger on it. Philosophers have come up with various theories through the ages. Perhaps it is everywhere, in rocks, or even in atoms [2]; perhaps it is a fundamental part of nature like space and time [3]; perhaps it is only an illusion and we only think we are conscious, so that there is no hard problem to solve [4]; perhaps it is an unanswerable question [5].

These theories and others may have their advantages, but some of them do not place consciousness in the brain, or do not explain its evolutionary cause.

---

[*]Correspondence author: Henry Grynnsten, Independent Researcher, Sweden. E-mail: grynnsten@hotmail.com

1. Chalmers, D: "Facing Up to the Problem of Consciousness", 1995. http://consc.net/papers/facing.html
2. i.e., panpsychism.
3. Hoffman, D. D: "The Origin of Time In Conscious Agents", 2014. http://cogsci.uci.edu/~ddhoff/HoffmanTime.pdf. See also David Chalmers in the TED Talk "How do you explain consciousness?", 2014. https://www.youtube.com/watch?v=uhRhtFFhNzQ
4. Daniel Dennett calls his view about consciousness "illusionism". See Dennett, D. C: "Illusionism as the Obvious Default Theory of Consciousness", 2016. https://ase.tufts.edu/cogstud/dennett/papers/illusionism.pdf
5. I.e new mysterianism, according to which the hard problem of consciousness is too hard for humans to solve.

Some scientists, and philosophers like John Searle with his biological naturalism [6], place consciousness in the head of the human being, and think that it can be explained by processes in the brain. This is the basis for the model presented here.

An idea of how consciousness could come about can be based on normal consciousness and the experience of déjà vu, which is a dislocation of normal consciousness. That almost everyone has experienced déjà vu indicates that it might be a key to understanding consciousness, since it affects consciousness.

Déjà vu is an experience that as many as 70 percent of all people have had [7], in some surveys 80, 90, or even 100 percent, though lower in others [7], while the rest maybe have forgotten or not been able to put a name on the phenomenon. Some are reluctant to admitting the experience, out of fear of being seen as abnormal [8].

A third experience, of heightened awareness, a counterpart to déjà vu, may be hypothesized. This would be much rarer, and may not even have a name, but will be briefly described below, since, if found, it could bolster the ideas here presented.

2. How Consciousness Arises

When we receive input from the senses, it is synchronized in the brain. We all seem to hear and see and smell etc. the world all at the same time. Our own experience tells us that different sense data are connected together in the brain, in one way or another. Let us call this the synchronization. The claim is not that this occurs in a specific region or site or by a specific mechanism in the brain, it is just shorthand for the way, whatever it is, in which disparate senses are synchronized for an experience of the world.

To summarize:

1) We have different senses that give us sense data;
2) We experience sense data not as separate, but as a total experience;
3) Therefore, in some way, sense data are synchronized; and
4) For convenience, in this article, this [i.e. point 3] is called the synchronization.

---

6. See for example Ford, J. B: "Aspects of John Searle's biological naturalism", 1999; "... according to biological naturalism, mental phenomena are higher-level physical features of the brain, but are totally caused by lower-level neurophysiological processes ...". https://minerva-access.unimelb.edu.au/handle/11343/114496
7. Brown, A. S: *The Deja Vu Experience. Essays in Cognitive Psychology.* Psychology Press, 2004. p. 34–37.
8. Brown, p. 30.

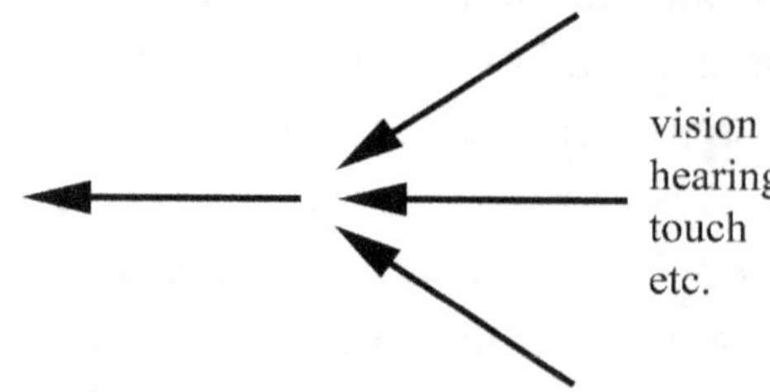

Fig. 1. Sense data are bound together in the brain

The synchronization – i.e. the connecting together of sense experience, however it occurs – might be the key to human conscious experience. The feeling of consciousness that humans experience could come into being by a doubling of the synchronization of the sense data, possibly by electrical and/or chemical impulses being doubled, making a reprise. If sense data are synchronized, which we can deduce from the fact that we see and hear etc. all at the same time, then it is possible that this synchronization, this binding together, which occurs once, could in some way be repeated.

In fact Robert Efron, looking into déjà vu, discovered that the temporal lobe receives incoming information twice, once directly into the left hemisphere of the brain and once by a detour via the right hemisphere [9].

Déjà vu can be defined as "any subjectively inappropriate impression of familiarity of a present experience with an undefined past" [10]. It is a phenomenon where we feel familiar with something we should not be familiar with. Normal memory follows from us having come across something before. So when we feel familiar with a present experience in déjà vu, that would mean that the brain has already become familiar with the present experience. This indicates that the feeling of familiarity might come after an unconscious primary experience, which in fact means that it is a double experience. This double experience means a doubling of the sense data.

This resembles the double perception explanation of déjà vu, which explains it by two perceptions in quick succession [11]. It also resembles dual process explanations of the phenomenon. Of four different types of disruption between the two separate cognitive processes proposed, one is caused by "an atypically long separation of two functions that are normally immediately contiguous" [12]. Already in 1897, E Parish suggested that an abnormal widening

---

9. Robert Efron's paper was "Temporal Perception, Aphasia and Déjà Vu", 1963,
   https://academic.oup.com/brain/article-abstract/86/3/403/260382?redirectedFrom=fulltext, described by
   Obringer, L A in "How Déjà Vu Works" at https://science.howstuffworks.com/science-vs-myth/deja-vu4.htm
10. Vernon M. Neppe quoted in Brown, A. S. P. 17.
11. Brown, p. 173 ff.
12. Ibid, p. 127–128.

"between sensation and perception" leads to déjà vu [13]. Alan S Brown also sees a similarity between this, the separated process idea, and double perception, as well as a neurological explanation that proposes that "information from the primary and secondary sensory pathways" reach the cortical processing centres with an unusual delay [14].

In 1895, Myers proposed a dual consciousness consisting of a subliminal and supraliminal self. The subliminal self records events continuously in the moment, while the supraliminal lags behind, but is the one we are consciously aware of [15].

All these ideas, while not talking about consciousness per se, seem to point in the same direction, of a doubling of sense data. You could call the first self-awareness, and the second self-consciousness, which, the two together, leads to the consciousness of humans.

By explaining consciousness as a phenomenon of memory, one gets around the problem of one specific thing being its source. It is not one thing, it is a succession of things. By being a combination of several things in time, no one single thing has to bear the burden of somehow giving rise to consciousness.

In normal consciousness, because of this repetition or doubling, you are just on the brink of feeling that you have experienced everything before. A little shift of normal consciousness, and the vivid sensation of déjà vu appears. Instead of the clear memory of déjà vu there is a feeling of being present in the here and now. This can be described as "I recognize myself" or "I am", "I am here" or "this is here and now". It is here that the "I" would come into being.

The result is, as David Hume puts it, that "The mind is a kind of theatre, where several perceptions successively make their appearance ... They are the successive perceptions only, that constitute the mind ..." [16].

This is not a homunculus argument, that there is a small man inside the brain who experiences this feeling. On the contrary, this model shows that there is no separate entity that is the self in the brain. The feeling of the self is in that sense an illusion. Consciousness clearly exists, but it does not indicate any separate entity or self in the brain. We are individuals with consciousness, but there is no self.

Just as animals, even primitive animals, can be shown to react to stimuli through the senses and thus feel in some way, in the same way a human can feel consciousness. There is no need for a homunculus or any mysticism.

---

13. Ibid, p. 132.
14. Ibid, p. 133.
15. Ibid, p. 134.
16. Hume, D: *A Treatise of Human Nature*, 1739–40. https://www.gutenberg.org/files/4705/4705-h/4705-h.htm

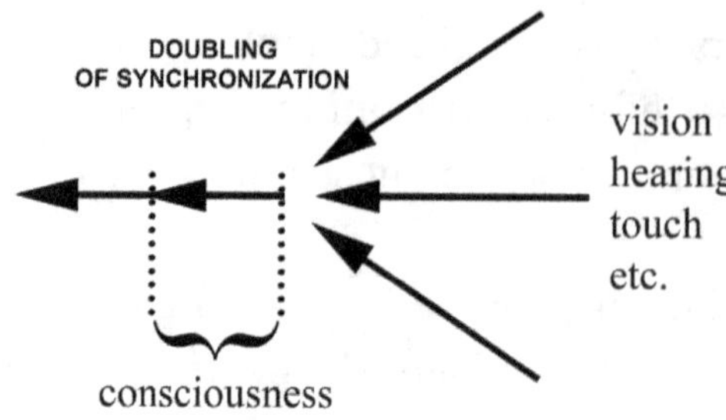

Fig. 2. The binding together of sense data can be repeated

There has long been an idea that déjà vu is caused by neurons firing out of sync. It is the contention of this article that it is when this doubling of the sense data is increased by a certain amount of time that we get a strong feeling that we have experienced everything before, a contention that has some antecedents in previous déjà vu research, as mentioned above. It is a feeling that is sometimes accompanied by depersonalization, a detachment of the self. The sense data in déjà vu begin to feel more like a memory than a direct experience, which shows that there has arisen a distance between the first and second binding together of sense data. Efron found that the brain registers the second piece of information as a memory because it had already been processed [17].

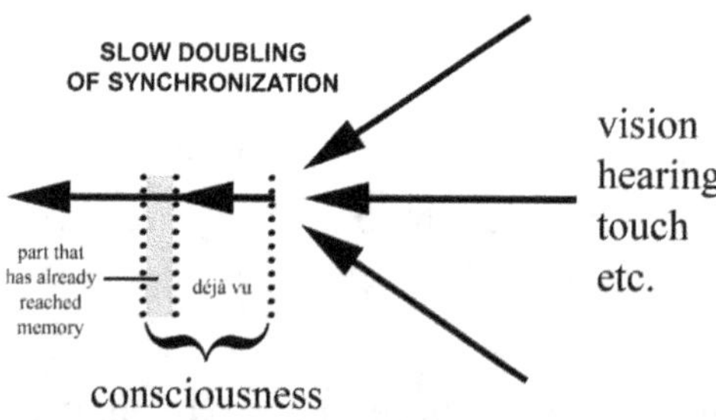

Fig. 3. Slow doubling of synchronization gives rise to déjà vu

If this is true, there might be a corresponding experience when the synchronization is faster than usual, when the distance between binding 1 and binding 2 decreases by an unusually large amount, a phenomenon, if found, that could bolster the explanation of déjà vu described here.

---

17. Obringer.

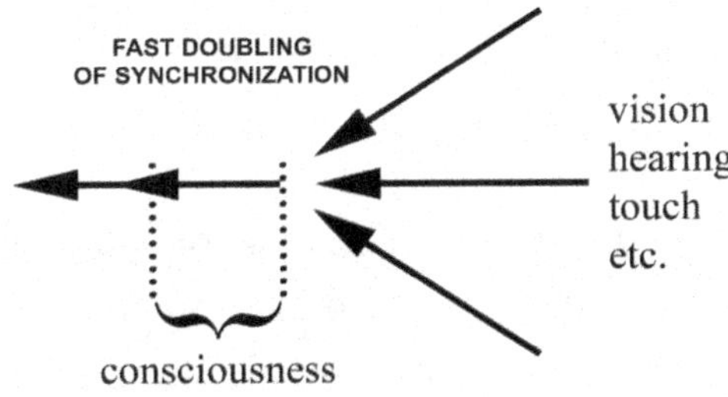

Fig. 4. Hypothesized fast doubling of synchronization

This could result in the opposite of the depersonalization of déjà vu, and be a state of heightened awareness of the self, if the supposition is true. It would be tempting to assume that this state is identical with jamais vu, the feeling of unfamiliarity with familiar places, objects or people, which is much rarer than déjà vu [18], but this might be a case of too easy parallelism and the explanation for the two phenomena could be two quite different mechanisms. However, déjà vu and jamais vu have been viewed by many researchers as "opposite ends of a familiarity dysfunction" [19].

To continue the supposition, the fast doubling of the synchronization might be a state that is just as distinct as déjà vu, but very different, and like it perhaps lasting only 10 to 30 seconds. This state may be described in the literature, but besides jamais vu it would also possibly not be identical with the "heightened awareness" described by meditators, since it would have such a short duration.

So the normal function of consciousness can be deduced from an experience close to normal consciousness, déjà vu, where synchronization is malfunctioning, that is experienced at some time by many people, perhaps most. If the second state could be demonstrated to exist, it would strengthen the argument here presented.

3. Why Consciousness Arose

Why do humans and only humans have [human] consciousness? What is it that humans do that no other animals do? Cognitive neuroscientist Merlin Donald writes that humans evolved "skilled rehearsal and explicit memory retrieval" [20], and this was used for language. According

---

18. Brown, p. 104.
19. Ibid, p. 106.
20. Donald, M: "The neurobiology of human consciousness: An evolutionary approach", 1995.
    https://www.sciencedirect.com/science/article/abs/pii/002839329500050D

to Donald, apes do not rehearse to improve performance [21], and they are the animals that are closest to humans cognitively. Explicit recall "depends heavily on having a symbolic system in place in the brain" [22].

Language is the most distinctive thing besides consciousness that differentiates humans from other animals, so you could assume that these two phenomena are connected in some way. If one presumes that they arose quite independently, one would need one mechanism to explain why consciousness arose, and another for why language arose, but in this article the assumption that they arose together and simultaneously is explored. This assumption does not seem to immediately encounter any logical obstacles.

It is thought that humans evolved cooking from around half a million to two million years ago. Over time this led to adaptations in the human body to cooked food, including a bigger brain [23]. This eventually led to consciousness, which made learning of language possible [24], and great advantages accrued to those who made better use of language. The "talkers" could outcompete their enemies and outsmart rivals, and store complex knowledge in their minds.

A creature with language as a tool has a clear advantage over one without language, since humans dominate all other species. The question is how far humanity would have come with just consciousness and no language at all, but it seems likely that once consciousness is in place, this will inevitably lead to better communication systems in some form.

Exactly how the doubling of sense data appeared, and if it first came about in one individual as a mutation, is not something that this article tries to explain. But through the doubling of the synchronization, by having increased consciousness of what we experienced, we were able to acquire language. By this means, we could sort out the interesting, new words or forms, and improve our language use. Soon only those who mastered language remained, all others having been outcompeted, or having disappeared for some other reason. There are now no human groups who cannot learn languages.

---

21. Donald, M: *A Mind So Rare. The Evolution of Human Consciousness.* W W Norton & Company 2002 [2002]. P. 142.
22. Ibid p. 163.
23. Wrangham, R: *Catching Fire. How Cooking Made Us Human.* Profile Books 2010 [2009].
24. An idea developed in Donald's *A Mind So Rare.* Donald argues that we can only learn consciously. He takes the example of reading; "The total conscious load imposed during the learning of advanced literacy skills is enormous and absorbs our attentional capacity for years." But then it becomes automatic and we no longer need to use enormous conscious capacity to read. [pp. 231– 232] Unconscious learning does happen, but for example learning during sleep is "extremely basic", writes Bahar Gholipour for Live Science; "Can You Learn Anything While You Sleep?", 2019. https://www.livescience.com/64920-how-learn-during-sleep.html.

Human ancestors in all likelihood had some forms of vocalized communication. Meerkats and other animals can probably change alarm calls depending on the situation [25]. Our close relatives the chimpanzees produce a range of vocal signals, but this communication is static, it does not accumulate over the generations and does not constitute a language. However, after humans achieved consciousness, our ancestors would have been able to consciously process these pre-existing vocalized signals, organize and elaborate them, extend their meaning, as well as to apply them to new situations. This elaboration would then become the new repertoire, which could be used without much reflection. But this new repertoire would then be subject to new conscious processing, and so on, until the stage of full, human language was reached.

Pidgin and creole languages show how quickly a language can come into being. Pidgin first develops when speakers of mutually unintelligible languages meet and have to communicate. Both the grammar and the vocabulary are very limited. Children of pidgin speakers then learn the language in the process called nativization, and develop creole, a full language with fully developed grammar and vocabulary [26]. Linguist John McWhorter says that "... of the languages extant today, the ones that most closely approximate the first language are creoles ..." [27].

This process can of course only be achieved by creatures with human consciousness – it may be unnecessary to point this out, but it has never been possible to teach any other animals human language. Animals communicate by static systems, while language is continuously changing. These are two entirely different modes of communication. Clearly, human communication requires mental flexibility, i.e. consciousness.

For gradualism, this would mean a slow evolution in tiny incremental steps from 10 to 20 to 30 to finally 100 per cent of modern human consciousness and language, say over hundreds of thousands or half a million years. If human ancestors had full consciousness at any point in the process, they would be able to quickly evolve full languages without intermediate steps.

If consciousness appears suddenly it must mean that those who are now conscious will be able to perform an operation similar to those who create pidgin languages. The early humans did not have full languages to work with, but they did have some kind of vocalized signals. These people, who were modern, conscious humans, must of course have had the same capacity to create a means of communication that is close to the pidgin languages as present-day humans. In fact, talking about deaf children without contact with sign language, a group of researchers write

25. Townsend, S W et al: "Flexible alarm calling in meerkats: the role of the social environment and predation urgency", 2012. https://academic.oup.com/beheco/article/23/6/1360/191395.
26. See, for example, Crystal, D: *The Cambridge Encyclopedia of Language*. Guild Publishing 1988 [1987], p. 334–337.
27. McWhorter, J H. *The Power of Babel. A Natural History of Language*. Perennial 2003 [2001], p. 301.

that "There are circumstances under which a small group of people can form a language apparently out of nothing" [28].

This developmental stage from pre-language signals to full language could thus have occurred very rapidly, only to reach a kind of ceiling – the ceiling that is exemplified in full human languages. And clearly there must be a reason for the existence of a ceiling in the complexity of language. Both the saltationist and the gradualist must explain why language no longer evolves [only changes].

At some point we reached this language plateau, since all human languages seem to be roughly equally complex [29]; they can all be translated into the others without too much loss of information.

From this time, we can surmise that language only changed in content, not in overall complexity [though local complexity of various parts of a language can and does increase and decrease], which indicates that consciousness also remained at the same level.

This would mean that language cannot develop far beyond current overall complexity [for general communication purposes] without a corresponding increase in consciousness, and vice versa [without going into if either language or consciousness can in practice reach a "higher" level]. If humans can develop a much more complex language with current consciousness, there would be examples of this, which there are not. If there were humans with much more advanced consciousness, they might use language at the level of complexity we see in any current or past language, but they would distinguish themselves in some noticeable way. Clearly there are not and have never been superhumans, that is humans that are to us as we are to other primates. Geniuses still fall within a kind of general range of human intelligence.

It is not a coincidence that children develop consciousness at the same time as they learn language. By 5 months, consciousness and memory are on their way, [30] and by 6 months, "most babies recognize the basic sounds of their native language". [31]

---

28. Meir, I et al: "Emerging Sign Languages", 2010.
http://sandlersignlab.haifa.ac.il/html/html_eng/pdf/EMERGING_SIGN_LANGUAGES.pdf
29. Linguist David Crystal writes that "... every culture which has been investigated, no matter how 'primitive' it may be in cultural terms, turns out to have a fully developed language ... Anthropologically speaking, the human race can be said to have evolved from primitive to civilized states, but there is no sign of language having gone through the same kind of evolution … There are no 'bronze age' or 'stone age' languages … All languages have a complex grammar: there may be relative simplicity in one respect [e.g. no word endings], but there seems always to be relative complexity in another [e.g. word position]." Crystal, p. 6.
30. Gabrielsen, P: "When Does Your Baby Become Conscious?", 2013.
https://www.sciencemag.org/news/2013/04/when-does-your-baby-become-conscious
31. The National Institute on Deafness and Other Communication Disorders [NIDCD] : "Speech and Language Developmental Milestones". https://www.nidcd.nih.gov/health/speech-and-language

"Perhaps language is even a necessary condition for consciousness" writes Psychology Today [32]. Or perhaps consciousness is a necessary condition for language.

This leads to the further observation that there are no zombies in the philosophical sense, since to learn a language you have to be conscious. I know that I am conscious, but how do I know that you are conscious? Because you use a language. Of course you can program a computer to give sophisticated answers to questions, but that is automation, not "using language" as humans do.

A person might talk during sleep, i.e. in an unconscious state, but he or she still learned language while conscious. The same goes for other exceptions: the individual must at some point before such an exception have been conscious for a considerable amount of time, the time that is required to learn a language.

It is also possible to date the appearance of consciousness: it would have happened conjointly with the development of language. McWhorter, for example, thinks that a mutation created a genetic instruction or predisposition for language, that it happened once, and possibly around 150,000 years ago [33]. Changing this supposed predisposition for language to consciousness makes the idea fit into this article. Crystal also is of the opinion that language "emerged within a relatively short space of time", and gives a possible date as late as 30,000 years ago [34].

It is thought that humans first started using abstract signs around 200,000 years ago. Language most likely developed before humans spread out from Africa around 60,000 years ago [35], which means that consciousness would also have arisen sometime between those dates, perhaps 70,000–100,000 years before present. Noam Chomsky talks about a sudden creative explosion around 75,000 years ago and connects it with language. "Well, what would have happened? Whatever happened ... would be some rewiring of the brain" [36].

Others, however, think that language could go as far back as 500,000 or 600,000 years before present [37]. That would also push back the appearance of consciousness to that time. An alternative to a quick development of language is a slow, gradual evolution over long time periods.

---

32. Haladjian, H. H: "Consciousness and Language", 2016. https://www.psychologytoday.com/us/blog/theory-consciousness/201608/consciousness-and-language

33. McWhorter, p. 8.

34. Crystal, p. 291

35. Noam Chomsky are among those who suggest an origin for language at about 70,000 to 100,000 years ago. Bolhuis J. J; Tattersall, I; Chomsky, N; Berwick, R. C: "How Could Language Have Evolved?", 2014. https://chomsky.info/20140826/

36. Chomsky, N. in "Grammar, Mind and Body – A Personal View", a talk in 2012 at "The 2011–2012 Dean's Lecture Series", at around the 1:13 mark. https://www.youtube.com/watch?v=wMQS3klG3N0

37. Dediu. D. and Levinson, S. C: "On the antiquity of language: the reinterpretation of Neandertal linguistic capacities and its consequences", 2013. https://www.frontiersin.org/articles/10.3389/fpsyg.2013.00397/full, and Quentin D Atkinson, University of Auckland, personal communication, 13 Oct. 2019.

The conclusion is that it is more likely that consciousness emerged suddenly rather than gradually, that it made it possible to construct language out of pre-existing vocal signals, and that this process very quickly led to a full language in a process similar to the one from pidgin to creole. This first language, also very quickly, reached a plateau of complexity that is the same as in all modern languages.

4. Animals and Awareness

You could make a distinction between animal awareness and human consciousness. If any animal had human-like consciousness, it would certainly be able to speak or at least show cognitive abilities on a par with humans in some way.

Therefore, it is reasonable to make a distinction between human consciousness and animal awareness. This is also useful as it can clarify what exactly is under discussion.

Awareness clearly varies between animals, from insects to primates. Some animals pass the mirror test, most do not. There is a gradation, beginning at an unknown, and perhaps unknowable, starting point, up to and including animals who pass the mirror test.

This means that awareness gradually developed during the evolution of animal life on Earth, leading to great diversity. But there was a clear break when humans became conscious, there is a qualitative difference between animal awareness and human consciousness. A great many species of animal have some form of awareness, but only humans have consciousness.

There seems to be a progression of what has been called consciousness from the "lowest" kinds of organism up to the "highest", i.e. the human. This has led a great number of people, both lay people and scientists as well as philosophers, to assume that this progression is more or less inevitable and will lead to ever higher levels of consciousness and intelligence. A further idea has been that life on other planets also has gone through this evolutionary progression, so that there will be aliens with super-intelligence.

However, human consciousness could be a rare occurrence in the universe, while awareness is common [since many animals seem to have it]. There is no evidence that human consciousness will continue advancing to higher levels. Instead, there could be a continued progression and diversification of awareness without consciousness in animal species. There is no reason to assume an automatic, progressive development of human consciousness, which appears to be a singular occurrence on Earth, but we know that awareness in animals has developed for hundreds of millions of years.

5. The Speed of Consciousness

By the model presented here it is even possible to arrive at what might be called the speed of consciousness. It is 300 milliseconds.

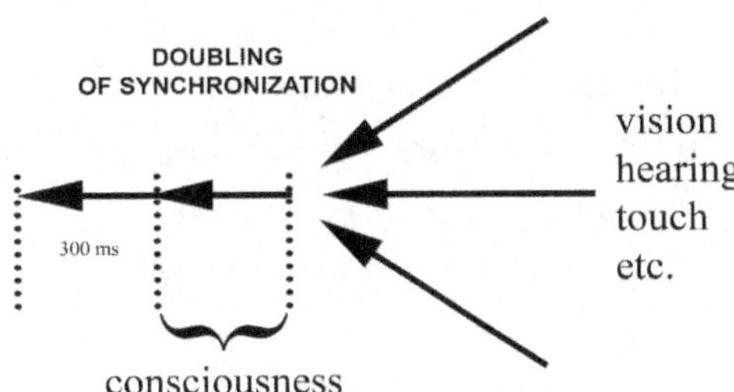

Fig. 5. The speed of the doubling of the synchronization

This is based on research where it has been found that the brain has decided to act before we are even conscious of it. In one experiment the human guinea pigs got electrodes attached to their scalps and were told to push a button at will; "it took a third of a second from the time the motor cortex initiated the instruction to act before the subject was conscious of the intention to act. In other words, the actual decision to act started the action sequence before the conscious decision to act" [38].

We cannot decide to act before we have received the sense data that there is anything to do in the first place. So the decision to act must come after the sense data has entered the brain, and, as demonstrated, consciousness catches up after the decision has been taken unconsciously, and that is after 300 milliseconds. In déjà vu, there is often a feeling of time slowing down [39]. This would agree with the model presented here of the increase of the time of the binding together of sense impressions during déjà vu.

The experiment also confirms the conclusion in this article that there is a kind of doubling of the synchronization of sense data, since clearly the decision to act repeats, first "unconsciously", then consciously.

6. Why Consciousness Is Static

A side effect of the pressure to learn language was that [human] consciousness spread, and the strange feeling we all carry around in our heads and seems to be so mysterious to us became the same for all humans.

---

38. Described in an article by Diamond, M: "Awareness Consciousness as Agent of Causation and Action", 2019.
   https://medium.com/the-philosophers-stone/awareness-consciousness-in-causation-and-the-flow-of-time-97caefcfa71f
39. Brown, p. 188.

For consciousness to spread from maybe a single individual there had to be an evolutionary pressure, and that pressure seems to have been language. However consciousness appeared, it was then used for language, which had such great advantages on an individual level that consciousness-language spread, and everyone who lacked this mental technology suffered great disadvantages.

Now both language and consciousness are essentially static [40]. The reason that language has been maintained, and not been lost by some human groups, must be that language fills an essential role, and for language learning to be possible there is a need for consciousness. Again, there is a great divide between static animal communication and human language, that never stays still and is always in flux.

It is probably hard in any society to find a mate if you lack language. You cannot easily communicate your intentions without language in a language-based society with complex rules. This means that there is constant pressure to learn language, so consciousness has to be maintained at a certain level.

But why has language and consciousness not evolved in complexity? Three possible explanations are:

1) There was and is no evolutionary advantage to even more complex language – we can communicate quite well inter- and intra-personally with the language we have, even develop philosophy and science.

2) Due to structural reasons, the human brain cannot develop to a level where it could give rise to the "higher" consciousness necessary to easily manage a much more complex language [i.e. a language so complex that no current human could learn it, just as no ape can learn a human language]. This does not necessarily mean, however, than *any* brain, with the right starting conditions, could *not* develop "higher consciousness" via an evolutionary process.

3) A much more *meaningfully* complex language is not possible to develop in the first place. For example, there are 3 cases in English, but 15 in Finnish, but Finnish is not "better" for communication than English. So a language with 1,000 cases would be more difficult [and take longer] to learn, but would probably not be better at communicating anything. Maybe language as a communication system simply cannot develop beyond a certain level – a level that we reached very quickly after we became conscious.

---

40. Chomsky, N. in "Grammar, Mind and Body – A Personal View", says about language development: "So nothing's happened for about 50,000 years", i.e. the last 50,000 years.

7. Conclusion

Through the mechanism of the doubling of the binding together or synchronization of sense data, [human] consciousness can be explained as a phenomenon that arises when sense data are repeated at a short interval, i.e. 300 milliseconds, which means that a memory element is involved; consciousness is a phenomenon of memory. When this distance is increased, one experiences déjà vu.

Consciousness appeared simultaneously with language, probably in a short span of time, and it is consciousness which makes human language possible. This means that consciousness can be surmised through the use of language: any creature using a full language would have consciousness, according to this model. There seems to be something missing in the description if one imagines a creature that had consciousness at a human level and no language or other flexible communication system. If language at a human level can be acquired without consciousness, the question is why no other animal can talk.

Answers to several questions can be begin to be answered with the help of this model of consciousness: how and when language arose, why language does not change in complexity, the future of consciousness and language, as well as other phenomena.

Consciousness does not result from not a single thing but from a succession of things in time. It is not a single structure, and it is a phenomenon of memory.

Acknowledgements: I would like thank the following for being helpful in various ways: Hawk Alfredson, Quentin D Atkinson, Tony Elgenstierna, Ahrvid Engholm, Peter Heft.

Received June 23, 2020; Accepted July 25, 2020

References

Offline resources:

Brown, A. S. [2004]: *The Deja Vu Experience. Essays in Cognitive Psychology*. Psychology Press, 2004.

Crystal, D. [1988]: *The Cambridge Encyclopedia of Language*. Guild Publishing 1988 [1987].

Donald, M. [2002]: *A Mind So Rare. The Evolution of Human Consciousness*. W W Norton & Company 2002 [2002].

McWhorter, J. H. [2003]: *The Power of Babel. A Natural History of Language*. Perennial 2003 [2001].

Wrangham, R. [2009]: *Catching Fire. How Cooking Made Us Human*. Profile Books 2010 [2009].

Online resources:
All resources accessed on 7 March 2020

Adachi et al [2003]: "Demographic and psychological features of déjà vu experiences in a nonclinical Japanese population". https://www.ncbi.nlm.nih.gov/pubmed/12695735

Bolhuis J. J.; Tattersall, I.; Chomsky, N.; Berwick, R. C. [2014]: "How Could Language Have Evolved?". https://chomsky.info/20140826/

Chalmers, D. [1995]: "Facing Up to the Problem of Consciousness". http://consc.net/papers/facing.html

Chalmers, D. [2014]: "How do you explain consciousness?". https://www.youtube.com/watch?v=uhRhtFFhNzQ

Chomsky, N. [2012]: "Grammar, Mind and Body – A Personal View". Talk at "The 2011–2012 Dean's Lecture Series". https://www.youtube.com/watch?v=wMQS3klG3N0

Dediu, D. & Levinson, S. D. [2013]: "On the antiquity of language: the reinterpretation of Neandertal linguistic capacities and its consequences". https://www.frontiersin.org/articles/10.3389/fpsyg.2013.00397/full

Dennett D. C. [2016]: "Illusionism as the Obvious Default Theory of Consciousness". https://ase.tufts.edu/cogstud/dennett/papers/illusionism.pdf

Diamond, M. [2019]: "Awareness Consciousness as Agent of Causation and Action". https://medium.com/the-philosophers-stone/awareness-consciousness-in-causation-and-the-flow-of-time-97caefcfa71f

Donald, M. [1995]: "The neurobiology of human consciousness: An evolutionary approach". https://www.sciencedirect.com/science/article/abs/pii/002839329500050D

Ford, J. B. [1999]: "Aspects of John Searle's biological naturalism". https://minerva-access.unimelb.edu.au/handle/11343/114496

Gabrielsen, P. [2013]: "When Does Your Baby Become Conscious?". https://www.sciencemag.org/news/2013/04/when-does-your-baby-become-conscious

Gholipour, B. [2019], "Can You Learn Anything While You Sleep?". https://www.livescience.com/64920-how-learn-during-sleep.html

Haladjian, H. H. [2016]: "Consciousness and Language". https://www.psychologytoday.com/us/blog/theory-consciousness/201608/consciousness-and-language

Hoffman D. D. [2014]: "The Origin of Time In Conscious Agents". http://cogsci.uci.edu/~ddhoff/HoffmanTime.pdf.

Hume, D. [1739–40]: *A Treatise of Human Nature*. https://www.gutenberg.org/files/4705/4705-h/4705-h.htm

Meir, I.; Sandler, W.; Padden, C.; Aronoff, M.: ""Emerging Sign Languages", in *The Oxford Handbook of Deaf Studies, Language, and Education*, Vol. 2, 2010. Edited by Marc Marschark and Patricia

Elizabeth Spencer.
http://sandlersignlab.haifa.ac.il/html/html_eng/pdf/EMERGING_SIGN_LANGUAGES.pdf

The National Institute on Deafness and Other Communication Disorders [NIDCD] [2017]: "Speech and
Language Developmental Milestones". https://www.nidcd.nih.gov/health/speech-and-language

Townsend, S. W.; Rasmussen, M.; Clutton-Brock, T.; Manser M. B: "Flexible alarm calling in meerkats:
the role of the social environment and predation urgency", 2012.
https://academic.oup.com/beheco/article/23/6/1360/191395

Obringer, L. A. [2006]: "How Déjà Vu Works" at https://science.howstuffworks.com/science-vs-
myth/deja-vu4.htm

Personal communication:

Atkinson, Q. D. [2019], University of Auckland, email 13 Oct. 2019.

Article

Role of Consciousness in Evolution of Human Society

Satinder S. Malik[*]

Abstract

Almost the entire world is in competition about economic progress, amassing wealth and energy resources. Thus, development of a sixth sense of humanity, common sense, to a certain level is required in order to change the direction of human society. The development of sixth sense lies in control of five senses. Suffering in life is also universal. In the causes of human suffering, lack of money has not been listed, and it is a matter of surprise that the world is busy uplifting human suffering by making economic progress. We inherit the planet and are blessed in the bountiful nature that needs to be preserved. Nature plays an unbiased role irrespective of faith or belief. Raising personal ethical standard also help the surroundings and social groups. The character of any group is as good as the vector sum total of characters of all the participants' combined. Therefore, tweaking of capitalisms characteristic is extremely important and that can be achieved by uplifting ethical standards to remove its suffering and ensure progress of human race.

Keywords: Consciousness, evolution, human society, sixth sense, common sense, capitalism.

Introduction

As we observe the world today, we find that the chief agenda on the minds of world leaders is economic growth, market economy and exploitation of earth's resources. The voices about the care for planet and of climate change are dismissed and brushed aside as the thought process of novices. The entire world seems to be in a race, a big competition about amassing wealth and resources. Corporations are maximizing profits without caring for impact on environment, on society or ant thought about worker welfare. The earth is being turned in to a concrete jungle. Exploitation of resources is so excessive that environment is becoming unstable and unpredictable. Global warming and sea level rise are staring to affect human life.

Instead of correcting ways, humans have started exploring other planets for human settlements. It is like leaving one in hand for the one in the bush. Even if humans migrate to such utopian future world, that will meet the same fate because unless we limit our requirements and mend our ways, the entire universe would not be sufficient. The businesses are creating skillful marketing campaigns with a lot of research on how to fool human brain in believing that what they are selling is their need and best product. The packaged food is laced with preservatives and addictive chemicals, dreams of getting richer and hopes of better sensual experiences are driving hordes of human population in hands of invisible entity called system. Reference [8] lists how some of the most famous people died. Drug overdose, heart failures, cancer, accident and suicide are seen above as most common killers in modern society. Do you think our social system is

---

[*]Correspondence author: Dr. Satinder S. Malik, Independent Researcher, India. E-mail: adventuressmalik@gmail.com

responsible for these deaths? Do we need to stop and think for a moment or carry on regardless, life is in a fast lane and we might miss out on something?

The religious code of social conduct of older times has been replaced by laws in every country, barring a few where they still follow religious laws. The religious laws are also debatable in the light of progress humans have made in various sciences. The respect for God has been replaced by a new system of governance. Visibly, there is practically no need for the concept of code, morality and ethics. Whether a system is good or bad would depend upon its provisions. If we go deeper in science we find hidden mathematics in nature, in the same manner the system we need to install above us also needs to confirm to mathematics of nature and that also needs to be compatible with the environment.

The Economic System

The most human societies are invariably ruled by the invisible capitalistic system today. We need to analyse whether it is the most beneficial way to organize our societies. Ancient legends in Greece speak of the early Pre-Socratics as traveling to India. Thales, Pythagoras, Empedocles, Democritus, and Plato were all fabled to have made the journey (although the legends are rarely given credibility). Commentators on early Greeks from around the first and second century passed BC on these legends. While these journeys may or may not have taken place, it is not unthinkable, for there were well established commercial routes between India and Greece along the Silk Road, protected by Persian king, as well as between ports on the Red Sea that linked Greece with India's thriving spice trade [7]. The trade in olden times was based on fair practices and did not uproot the society from their traditional homes.

While the socialism has failed, capitalism has also led to increase in rich-poor divide, generation of waste, uprooting families and destroying social support system, urbanization, harmful working conditions and pollution to name a few. Globalization is a further development of capitalism. Multinational corporations [1] are accused of social injustice, mismanagement of natural resources, and ecological damage and unfair working conditions as well as lack of concern for environment. These corporations, which were previously restricted to commercial activities, are increasingly influencing political decisions. Many think that there is a threat of corporations ruling the world (at least having major influence and say) because they are gaining power because of globalization. The goods are moved from one corner of the globe to another just because they are economically cheaper. The costs to the environment and society are intangibles and therefore not considered.

During the most recent period of rapid growth in global trade and investment, from 1960 to 1998, inequality worsened both internationally and within countries. The UN Development Program reports that the richest 20 percent of the world's population consume 86 percent of the world's resources, while the poorest 80 percent consume just 14 percent. Globalization is also leading to the incursion of communicable diseases. Covid-19 is the latest example which has exploded to a pandemic. Globalization [2] might stifle competition if global businesses with dominant brands and superior technologies take charge of key markets, be it telecommunications, motor vehicles and so on. A recent book on Interdependent capitalism has put forward the issue of lack of social system built around inclusive fitness. The people move from genetically related, genuine caretaking relatives to complete strangers.

Greed- Is it a Natural Human Behaviour?

Liberals tend to view capitalism as an expression of natural human behavior that has been in evidence for millennia (of trade) and the most beneficial way of promoting human well being [3]. They tend to see capitalism as originating in trade and commerce and freeing people to exercise their entrepreneurial natures.

The above statement is at the best vague because it is not dependent on logical argument but on a premise e.g. it has been happening like this since millennia. Also, the assumption that it is the most beneficial way of promoting human wellbeing is a gross overstatement. Capitalism aims to maximize the profit (in the garb of value). When the profits are threatened, corporations implement desperate measures to secure markets, tax heavens and cheap labour with little regards to safety in working conditions. False propaganda and clever marketing campaign further deceive the gullible public. Doctors write so many additional tests and procedures that patients have to buy health insurances. Politicians have to accept proposals and make policies that support businesses; in turn, businesses support them and finance their elections. For instance, a particular steel plant also had a thermal power plant for electricity generation. There was a government policy in vogue which allowed electric supply to heavy industries at cheaper rates. So this company bought electricity from the government at cheaper rates and sold thermal plant generated excess electricity back to the government at higher rates. This appears to be a smart play, but it was perfectly legal. There are many such loopholes that escape the scrutiny of the public eye.

Therefore, capitalism is a greed-driven system which is governing most societies. Sometimes it becomes difficult for the ethical employees to bear the burden of corporate lies. Others become part of the system. It has technically become the new religion. Businesses heavily depended on manual labour earlier, which was less productive, safety conscious and sensitive. New automated systems were built and in companies like Foxconn and they replaced the labour and provide very high efficiency and increased rate of production. Implementing artificial intelligence (AI) is going to replace middle level human recourses, which may have created problems and embarrassment for corporations due to their ethical thinking. The idea behind implementing such technology is not to help the society but to feed the greed.

In socialism [5], greed shifts from productivity into consumption. Without property rights or opportunities for profit, men quickly descend into mutually destructive envy. Our base instincts betray us. Output plummets. When we see someone slacking and still taking - we produce less. When we see others hoarding – we snatch more too.

The question is, are humans traditionally greedy? Or it is a misinformation campaign to justify accumulation of inappropriate levels of wealth. How trading has risen from being considered as lowly activity to the epitome of shining success? Is the activity that has transformed or have humans have lowered their standards? Is it possible that human beings can evolve and be free of so called inherent flaw of greed and accumulation? Can intellect and consciousness play a key role in this evolution? Can humans' base their conduct on this planet using common principles which are in synchronization with our nature and environment?

Laws of Nature in Universe

The God does not create any agency or make one to do any action. He never tells anyone, do this or do that. He does not ensure the fruit of actions. Innate laws of nature ensure everything. This is the meaning couplet from Geeta Chapter 5 verse no 14.

न कर्तृत्वं न कर्माणि लोकस्य सृजति प्रभुः।
न कर्मफलसंयोगं स्वभावस्तु प्रवर्तते।।5.14।।

The God is same for all humanity whether one believes him or not. God created the universe and life in universe, he knows everything; it is the humans who feel him using their sixth sense and call him various names. Various names do not alter the interpretation of the God. Dimension of consciousness is part of such nature and sixth sense is to perceive that dimension. In that dimension also, there is a spiritual organization which controls space races and humans and can prove to be helpful by giving ideas, messages and motivation. Therefore, the Supreme God is Parbrahma (beyond universe) and without attributes of Sat, Rajas and Tamas. He does not interfere with human actions or their results but there is a complete divine organization under him which is also part of nature. It is out of purview of science because of scientific apparatus are limited to five senses.

Div is the word root for Divinity, Dev or Devta meaning one of the Gods or member of the alien species. The universe is populated with innate consciousness (Kshar Brhama) and also with souls (Akshar Brahma). All entities of the universe have three attributes Sat, Rajas and Tamas. These manifest as truth/purity, control and binding forces. All entities with a soul have a free will and as per their motivation can develop in to Sat, Rajas and Tamas. They manifest as good, neutral and bad.

The space races Rudra, Vishavdeva, Dev, Pitra, Daitya, Danav, Yaksha, Gandharva, Rakshash, , Supernas, Kinnars, Bhuta, Preta and Pisacha etc. Bhuta, Preta, Pishachas and Pitra (manes) are human transformations. All these space races are arranged in following organizational tree. The naming is in Sanskrit, as per their characteristics.

Consciousness of Vishwadevas empowers the systems of stars and planets. Their offspring's with humans are popularly called star children. Successive lines of kings after Devas were given knowledge of the secret yogic techniques. Raj Yoga is a technique which was passed down to the line of kings so that their consciousness gets evolved and that they can rule as per Raj Dharma (duties of the King). Without such a high level of consciousness when modern human beings reach a position of power, a few of them tend to become corrupt, arrogant, cling to power and cause atrocities. Many civilizations of the past were also formed initially by alien beings. Danvas, Daityas, Naga, Supernas, Gandharvas and Kinners all have been instrumental in establishment of animal and human kingdom on Earth. In many cases Asurs, Daityas and Danvas were found acting detrimental to the plan of Brahma and were removed from earth against their wishes. These alien being still keep influencing human race with their ideas.

If someone from human race gets contacted by any of the above space races, it doesn't mean it is for good purpose (Devas) only. The Asuras, Danvas and Daityas may also influence the course of human thinking away from the path of consciousness to delay the progress of human race. They can also give knowledge about magic, sorcery and wizardry so that humans get dependent on it. All these arts are also part of the dimension of consciousness but can be considered as negative traits if used against others or for personal benefits. Probably, this is the reason Europeans are against this dimension because they may have had bad experiences with witches and wizards. Therefore, anyone who calls himself an angel may not necessarily an angel. Whether he is angel or not would be determined by the knowledge he imparts. God doesn't want his creation to be afraid of him or he doesn't believe in helping only believe or he doesn't want to wage war in his name. Such mischiefs are created by Asurs, Danvas and Daityas only for gaining control over humans and they always look for greedy, ambitious and ignorant people. Their methodology is to infuse weak willed people with the ideas, in their dreams and controlling them by controlling their senses. There is also no need to worry about fires of hell. If the body is left here on earth, sensors on skin allow you to feel hot or cold, without your body you cannot feel fire. There is no hell. If you are not learning or bad with your Karma, you may spend time analyzing, descend to lower creation to learn again or may become part of life on other evolving planets where you may get stuck for many thousand years.

Consciousness and Sixth Sense

 If we could connect science and consciousness, most problems would get solved. Although, such a dream is not a far from reality but basic limitations remain. One major basic limitation is that of sense perception. A total of approximately 84% population on earth are religious and believe in some form of supernatural power or super consciousness. That is a clear majority of five out of six people. However, these people are also not able to fully believe, trust or realise the true nature of consciousness because of misconceptions created by science. Science is a great subject when it comes to an enquiry about material objects. Humans assess other objects or phenomena in environment, based on the receptors which are in the sense organs. The senses of sight, sound, smell, taste and touch give an input which is interpreted by mind and further processed by intellect to draw useful inferences. All the implements of science in laboratories are an extension of our senses; they widen the bandwidth and scope of the senses and the obtained results are again interpreted by same senses. Based only on the deductions using these senses, scientists are unable to explore any other dimensions such as of the consciousness. Such higher dimension is able to be perceived by humans by a sense which we refer to as the sixth sense. Intuition, telepathy and premonition are some phenomena which are explained by such an assumption.

The sixth sense is more prominent in females as compared to males and animals are also endowed with this sense. The barking of dogs prior to an earthquake or knowledge about death of close family members by whales even if they are thousands of miles apart are some examples. Perceptive powers [4] like telepathy, sense of direction and premonition seem better developed in species like dogs than they are in humans. They also occur in the humans in traditional cultures than in humans of the modern industrial world. We may have lost some of these abilities because we no longer need them: telephones and television have superseded telepathy; maps and global positioning systems have replaced the sense of direction. And perceptiveness is not cultivated in our educational system. Indeed, the existence of unexplained powers is not only ignored but often denied.

By the laws of our design, we mainly have six senses. It is the development of sixth sense to a certain level which is known as common sense. The development of sixth sense lies in control of five senses. Human mind is a limited resource and its attention needs to be limited to other senses to ensure development of sixth sense. It is true today, and it was true in the past because it has been tested in tried.

It was the effect of evolved consciousness that a large number of kings in India used to retire in time and follow Sanyas Ashram (life of austerities away from family). Many kings had even donated their entire kingdoms for charity and become monks. Buddha and Mahavira, who founded Buddhism and Jainism, were both princes. King Ashoka left his kingdom and became a monk. King Harshavardhana of the 7th century grew up in Shaivism but he championed Jainism, Buddhism and all traditions of Hinduism. He donated his kingdom and everything else including his worn cloths and had to borrow a piece of cloth to continue his journey as a monk. Bhartṛhari of Ujjain was a king before renouncing the throne to become a yogī. Gopīcand [6] was the son of Queen Mayanāmatī of Bengal, who became a Nath Yogi. Mahabali and Bharta were two kings who were brothers who became Jinas (Jain adepts) and they were sons of King Rishabhnatha who became the first Tirthankar (Jain Sidha). King Vikramaditya (Chandragupta Maurya) followed Jainism and gave ancient India the Vikrami calendar.

Helena P Blavatsky [8] stated that at this time of life she began to experience visions in which she encountered a "Mysterious Indian" man, and that in later life she would meet this man in the flesh. The "mysterious Indian" who had appeared in her childhood visions, a Hindu whom she referred to as the Master Morya and that he had a special mission for her, and that she must travel to Tibet. The Theosophical Society influenced the growth of Indian national consciousness, with prominent figures in the Indian independence movement, among them Mohandas Gandhi and Jawaharlal Nehru, being inspired by Theosophy to study their own national heritage. Blavatsky believed that Indian religion offered answers to problems then facing Westerners; in particular, she believed that Indian religion contained an evolutionary cosmology which complemented Darwinian evolutionary theory, and that the Indian doctrine of reincarnation met many of the moral qualms surrounding vicarious atonement and eternal damnation that preoccupied 19th-century Westerners.

Jesus said that it would be hard for the rich to enter the Kingdom (Matt. 19.23). In Padua, Italy, in 1304, on the wall of a church, painter Giotto made a fresco 'Jesus and Money Lenders'. Jesus and his disciples had travelled to Jerusalem for Passover, where Jesus expels the merchants and moneylenders from the Temple, accusing them of turning the temple into "a den of thieves" through their commercial activities. It was an idea that by then had already been taken root in Europe, that good spiritual life and trade and money are sworn enemies.

Adam Smith could change the course of thinking in regard to the trade and commerce and pave way for future capitalism. The individual virtuous traits are mostly common in all religions, though their priorities may slightly be different. Practice of Dharma is for everyone. Dharma is set of principles (thinking and action). These principles would same for everyone in that role e.g. as parent, child, brother, sister, husband, wife, judge, administrator, service provider etc. Action (Karma) is individuals' duty, and he has the right of acting but not on the result because result is influenced by so many other factors not within one's control. The act of thinking is sacred and action must be preceded by right thought. The thought is of two types, the one which generates spontaneously and the one which is directed by intellect. The one which comes spontaneously needs to be filtered and the unfit ones need to be discarded. The thoughts which are processed further become base for the directed thought. The principles of thinking and action are to be based on the following values.

(i)   **Satya (Truth, Reality).** Being truthful to self and others, close to reality and in pursuit of reality. This also includes efforts made to stay out of the veil of Maya (illusions). The various illusions are caused due to modern audio-visual media, marketing campaigns, fake science etc. Honesty is also a derivative of the truth.

नास्ति सत्यसमो धर्मो न सत्याद्विद्यते परम् ।
न हि तीव्रतरं किञ्चिदनृतादिह विद्यते ॥

It means that there is no Dharma (code of ethics) equivalent to truth. There is nothing beyond truth. Nothing is known as more damaging then falsehood or unreality.

(ii) **Ahinsa.** Not harming anyone without a reason, peaceful nature and tolerance.

अहिंसा सर्वभूतेभ्यः संविभागश्च भागशः ।

दमस्त्यागो धृतिः सत्यं भवत्यवभृताय ते ॥18॥

Mahabharata mentions that one should not harm any living being, everyone should be given what is due to them and one should exercise control of senses, patience, practice of truthfulness, such practices are gratifying and uplifting. Live and let live is another expression for Ahimsa. Forgiveness is also a derivative of this. It allows a time of contemplation for a person making a mistake and a chance to correct himself. Mistakes are described as of two types. One are those which are done unknowingly and second are those which are done knowingly under anger, temptation, fear, jealousy, hatred etc. Forgiveness is to exercise for the first type and atonement for the second type. However, one should keep in mind that this is for the evolved society and coercion and proportionate force is required to establish order in the society. To restore Dharma, even the path of war is justified.

(ii)   **Gratitude.** Thankfulness for the help received. Similarly, the person helping must forget the help which he has given to others, help should not be given with any expectation or result in mind.

(iii) **Tapas.** Tapas stands for Perseverance, Contemplation, self-discipline, austerities and patience. Tapas are personal performances in spiritual, metal and physical plane. It helps a student to achieve his aim of gaining knowledge and being successful in his studies and similarly all others in their profession.

(iv) **Paramarth**. It is the ultimate cause of serving others, charity, kindness, benevolence, magnanimity etc. When a person is young, he is bothered about his survival and so he is sva-arthi (self-ish), when he grows up supported by his parents and society, he must be aware that he too has to support the society. This is paramarth (in service of others). This is what needs to be promoted in society. Heroism lies in serving others and not collecting for oneself.

The Need for Suffering

Suffering is common in all parts of the world. No one is in a state of bliss all the time. It appears to be applicable to all alike, rich and poor, strong and weak, old and young irrespective of caste, creed, sex or race. There are two aspects to suffering. One is that suffering is due to duality (both side limits of sensory perceptions) and their interpretation by human mind, for example, hot-cold, pretty-ugly, sweat-bitter, love-hate etc. The second aspect is that suffering is part of the design of the environment on Earth, necessitating humans to think and evolve. It is the necessity which is mother of inventions. The suffering has also been analysed at 3 levels in Sankhya philosophy. Tthese 3 levels are Adhi-bhautik (physical), Adhi-atmic (cognitive) and Adhi-daivik (divine). The divine order and will in form of sum of past Karma, effect of planets, blessings of Pitra (ancestors) can represent in form of spiritual suffering and subsequently in physical dimension. **The universal experience of suffering forces humans to find solutions for its removal, and that ultimately leads to realisation of the truth** (about Soul). The suffering is to be taken like a challenge, a riddle for human beings to understand reality. Therefore, suffering would remain till the time humans need to evolve and this issue is not likely to get solved with economic growth. It will be solved with learning. The aim of learning is Purushartha that has four components of Dharma, Artha, Kama and Moksha. Dharma applies to rest three; it is a code as described above. Artha means basic material wherewithal to fulfill (Kama) desires. The life needs to be lived practically fulfilling all desires but as per Dharma. So the paper is not trying to preach asceticism but moderation of wants.

In pursuit of happiness, the mind attaches to what it considers as a comfort causing experience, and it tends to stay away from sorrow and painful experiences. Maharishi Patanjali described working of mind in Patanjali Yogsutras. He also describes the factors which cause inflictions on mind.

अविद्यास्मितारागद्वेषाभिनिवेशाः क्लेशाः ॥२.३॥

Avidya – ignorance, Asmita – egoism, Raga – attachment, Dvesha – aversion (Hatred, Jealousy), Abhiniveshah - clinging to life, Kleshaah - causes of suffering.

A lack of insight (avidya) is the source of most inflictions of mind. Therefore, it is first and foremost duty of a human being to remove avidya by seeking knowledge and drawing inferences to gain vidya by reasoning and logical thought process. Avidya is of two types Sanskara born and sense born (Paper: Evolution of consciousness).

Ego can inflate very fast in gratifying conditions. False ego is generated when one starts associating himself with body and environment which needs to be avoided. Association with objects which are ephemeral, temporary and not under one's control would cause inflictions on mind during their separation. The notion that pain and suffering are caused by external circumstances is referred to as aversion (dvesha). Abhinivesha can cause anxiety. Anxiety can arise spontaneously in particular conditions and can even dominate individual existence. All the above causes of suffering need to be nipped in the bud.

There are some misconceptions which arise due to the illusions we have created for the mind. The acts of procreation are deceivingly well marketed through movies, TV series and portrayed as a new found freedom which is facilitated by harmful world of contraceptives. Availability of porn magazines and audio visual media engage attention of the mind and cause vrittis. Vrittis are whirlpools of thoughts in peaceful lake of human mind. Modern food habits are shaped by the extraordinary science of addictive junk foods.

Kama (desires), Krodha (anger), Madya (addictive substances), Lobha (greed), Moha (attachment) are all root cause for deviating from natural human behaviour. Marketing gimmicks create strong wants and desire and a fake value to materials. When the strong waves of a want, desire or thought persist in the mind, they become stronger and take shape of whirlpools or Vrittis affecting deeper intellect and also causing sanskara (long term habit). Falling in love with an object or a person is a vritti. Similarly, depression is also a vritti. Patanjali has also described a solution to overcome vrittis.

ध्यान हेयाः तद्वृत्तयः ॥११॥

Meditation on overcoming the vritti eliminates such misconceptions that arise from the vritti. **In the causes of human suffering listed above money is not even listed and it's a matter of surprise that the world is busy uplifting human suffering by making economic progress.** The basic needs of food, shelter and clothing need to be met. How much is too much can decided by personal capability. For example, how big a house one needs? Given a choice majority would prefer white house but our need a house should be limited by our capability to maintain it personally. Due to lack of understanding in causes of suffering and attachment to comforts, human beings are pursuing accumulation of material resources. It is a wrong notion that lasting happiness would be ensured when we have everything. Due to this approach true happiness has been largely elusive.

Effect of Individual Consciousness in human groups

What we need is individual common sense and beyond common sense for general public. For administrators, judiciary and policy-makers higher senses needs to be enhanced. People with good leadership ability who take up society's cause, nation's cause or any other common cause for world display this trait. Abolition of slavery, un-touchability, religious harmony, creation of world bodies like United Nations Organisation are some of these actions which far greater than any individual cause and are applicable to entire humanity. Humanity when divided over common aim of human existence is like domains in a non magnetized iron, whereas humanity united by common cause is like magnetized iron with all domains aligned.

(a)   (b)

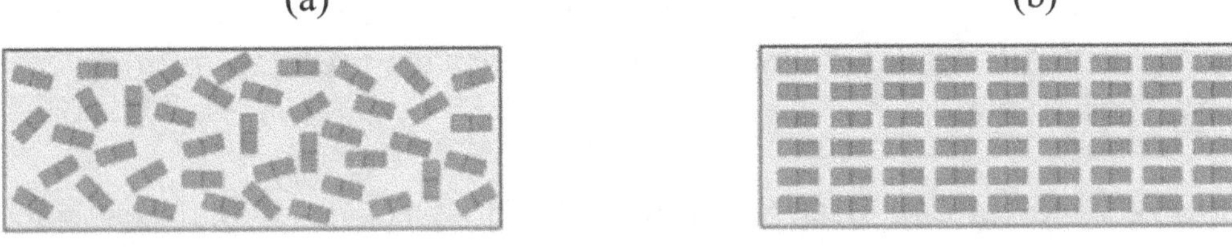

(a) Humanity without common understating of our nature and (b) Humanity with common understating of our nature

The vector sum of common sense and sixth sense in any human group be it a society, state or nation would not only define its character but also dictate its progress in various fields e.g science, arts, culture, economy etc. Such evolved consciousness in that forum, society, state or national identity would also reflect like an individual character (which is in fact vector sum of individuals' consciousness). For example, the moral values, helping nature, straight forwardness, honesty, ability to withstand adverse situations and order in society would be reflected by the behaviour and conduct of its people. Countries like Japan, Bhutan, Norway, Sweden, Finland and India on many different occasions have shown tremendous character with maturity, responsibility and understanding.

Intellectual Property

A way to understand a society is to look at the objects it produces. This is an idea deeply rooted in materialism for the study of anthropology and sociology. Though it is stated that countries establish intellectual property laws to foster creativity but these are actually more hindrance to creativity. These laws are for more the inventors to reap the benefits of their ingenuity.

ज्ञातिभिर्वण्टयते नैव चोरेणापि न नीयते ।
दाने नैव क्षयं याति विद्यारत्नं महाधनम् ॥

Knowledge is a great possession. It cannot be bisected, cannot be stolen, and by donating it further multiplies. The first line is in context for individual possession, but the last part gives the underlined summary that **by donation knowledge further increases**.

In 1710, The Statute of Monopolies changed that by allowing the author or inventor to retain their ownership rights [10]. Monopolies, in the form of government-sanctioned guilds, were no longer granted. The law also guaranteed the inventor, a 14 year period during which he had the exclusive right to govern how his invention was used.

In 1883, the Paris Convention ensured an international agreement through which inventors could protect their innovations even if they were being used in other countries. Writers came together in 1886 for the Berne Convention, which led to protection on an international level for all forms of written expression as well as songs, drawings, operas, sculptures, paintings and more. Trademarks began to gain wider protection in 1891 with the Madrid Agreement while the offices created by the Paris and Berne Convention eventually combined to become the United International Bureau for the Protection of Intellectual Property, the precursor of today's World Intellectual Property Organization.

Intellectual property by definition itself is intangible [9]. It is one of the most important structuring systems in modern society. It is the concern of many industries such as aerospace, architecture, pharmaceuticals, media and entertainment. Objects like the Lego brick, the Barbie doll, and the Coca-Cola bottle are heavily dependent on trademark protection. The value in these objects changed the IP system, as the companies controlling them had a hand in influencing the developments of law.

The world is progressing in a material dimension, and ideas are multiplying. We need to understand the nature of the human body and we will realise from where the ideas come and are they really an individual's property? Telepathy is a known communication process using human mind. The brain waves are able to transcend great distances. Every human being also has a subtle energy body and aura of this energy body can extend in a radius of hundreds of feet from an individual. Therefore, when we interact with people, the minds start communicating without conscious brain knowing about it. The ideas are floating in space above all of us; a person receptive to that particular subject is more likely to receive the idea. In many cases, scientists have received solutions of problems they were contemplating upon, in their dreams by the blessings of the divine. Therefore, whereas an individual must be celebrated and acknowledged about his contribution in that area, he may be additionally given a grant, a prize or limited time duration for exclusive usage depending on its nature but not exceeding five years. It is natural that if he had not discovered that somebody else would have done that slightly later.

The second argument against intellectual property rights is the fact that, the way there are raw materials for a product, there are raw materials for an idea too. There raw materials for an idea are underlining concepts, education, influence of society and a series of human innovations starting from fire, wheel, agriculture, fundamental sciences etc, all are a launch pad for him and he is just processing this information and drawing new inferences. Is anyone paying IPR for the decimal number system, philosophy, religion, ayurveda, yoga, etc? People who discovered or enhanced the knowledge in this world, left these as a gift to mankind and greedy industrialists are fighting for their meager contributions to the great knowledge of humanity for money. How cheap does it sound? We need to remove all barriers in distribution of knowledge and make it free flowing so that, it multiplies.

Assessment of Impact of technology on human Society

The body and mind of human beings was designed for environment of earth keeping in mind length of day, seasons, gravity, availability of water, food and other resources. Farther we proceed from the nature; more alienated we become from our true self and lose our health, happiness and harmony in that bargain. Just to look at the capability of human mind, thousands of Sanskrit Shalokas (couplets) were used to be remembered and passed down to generation in the form of Shrutis. When the palm leaf writing and subsequently books were invented, this power was lost or forgotten because there was no need of that effort. We are not in the habit of remembering the small figures and numbers. We have lost the accurate sense of date and time, for small calculation we need calculator and computers. We could walk thousands of miles or use natural resources for transportation which were pollution free, environmentally friendly, and those didn't give us jet leg. Our individual capability and knowledge is so depleted that most of us would not be able to survive outside the artificial city environments.

The developments in science and technology sound great for comfort and material resources. The need for comfort is also leading us to weakness and less perception. Various radiations have poor health impact on our body. Technologies like 5g are being pushed without any trials of its impacts on public health. Artificial intelligence is going to make us further dependent on material resources. Robots will do all the work for high and mighty, making them further incapable of doing any significant work. We must remember that the challenges are needed to be faced, not removed for the purpose of human evolution, growth and progress.

Reforms in Current Economic System

All the world bodies in United Nations must accept the problem. The joint body must identify the areas of concern and address them. The following measures are recommended.

(a)   It needs to be understood that economic progress only is not the real progress. Progress of human values is the real progress, and it will help uplift the moral and ethical standards of human society.

(b)   Working out an agreeable understanding about aim of human existence on earth and objective for achieving this aim, needs to done in a world forum. This is irrespective of religious beliefs, faiths and dogmas. Such an understanding needs to be progressive, logical and of reason.

(c)   An extensive research on causes of human suffering and adopting technologies and methodologies in society to prevent pre-mature deaths.

(d)   Decentralization of Industrial base. A stable society without much uprooting, smaller industries. Industries location based on availability of resources.

(e)   A general guideline about amount of percentage of profit making, display of cost of making and price of the product promoting transparency in businesses.

(f)   Awareness and regulation of marketing campaigns for their content and intent. Regulation agencies to access addictive nature of products and technologies.

(g)   Abrogation or modification of Intellectual Property rights.

(h)   Educational reforms to uplift ethical standards.

(i)   Continuous identification and elimination of non-contributing professions which are present due to complex nature of rules such as lawyers, middlemen, agents etc.

Conclusion

The entire world is in competition about economic progress, amassing wealth and energy resources. Corporations are maximizing profits without caring for impact on environment, on society or any thought about worker welfare. Most celebrated and successful people are also unable to complete their lives. Drug overdose, heart failures, cancer, accident and suicide are seen above as most common killers in modern society. The failures of modern society cannot be overstated.

Humans mainly have six senses. Development of sixth sense to a certain level is required, and it is known as common sense. The development of sixth sense lies in control of five senses. Suffering in life is also universal. In the causes of human suffering listed above, money is not even listed and it's a matter of surprise that the world is busy uplifting human suffering by making economic progress. We inherit the planet and are blessed in the bountiful nature that needs to be preserved. Nature plays an unbiased role irrespective of faith or belief. Raising personal ethical standard also help the surroundings and social groups. The character of any group is as good as the vector sum total of characters of all the participants' combined. The greed

(Published in Journal of Consciousness Exploration & Research| August 2020 | Volume 11 | Issue 5 | pp. 454-467)
Malik, S. S., *Role of Consciousness in Evolution of Human Society*

is sole motive of Capitalism, while it may be good motivation for humans to get them out of laziness and be good entrepreneurs but it also need to be controlled, lest it become monstrous and finish the human race itself. **Therefore, tweaking of capitalisms characteristics is extremely important and that can be achieved by uplifting ethical standards to remove its sufferings and to ensure progress of human race.**

Received April 28, 2020; Accepted July 5, 2020

References

[1] https://www.forbes.com/sites/mikecollins/2015/05/06/the-pros-and-cons-of-globalization/#1cc50385ccce

[2] https://listverse.com/2012/01/16/top-10-disadvantages-to-capitalism/

[3] https://en.wikipedia.org/wiki/Capitalism

[4] https://www.ru.org/index.php/consciousness/263-the-unexplained-powers-of-animals

[5] https://www.forbes.com/sites/billflax/2012/01/31/was-jesus-a-socialist-capitalist-or-something-else/#22b0a2a07324

[6] Nath Sampradaya Koninklijke Brill NV, Leiden, 2011

[7] http://grahamhancock.com/phorum/read.php?1,103556,103883

[8] https://www.usmagazine.com/celebrity-news/pictures/most-shocking-celebrity-deaths-of-all-time/41272/

[9] https://www.wipo.int/wipo_magazine/en/2019/04/article_0007.html

[10] https://txpatentattorney.com/blog/the-history-of-intellectual-property/

Article

Panpsychism as an Observational Science

Gregory L. Matloff[*]

Physics Dept., New York City College of Technology, CUNY, Brooklyn, NY, USA

Abstract

The metaphysical concept of panpsychism defines a field of proto-consciousness that is present at all levels in the universe. An observational indication that this concept might be correct is self-organization on all levels from the molecular to the galactic. In 2011, an investigation into the validity of author Olaf Stapledon's concept that a portion of stellar motion is volitional led to a consideration of an observational stellar kinematics anomaly dubbed Parenago's Discontinuity. Data available at the time indicated that cooler, less massive stars (such as the Sun) revolve a bit faster around the center of the Milky Way galaxy than hotter, more massive stars, at least in a sphere with a radius of ~260 light years centered on the Sun. The spectral signature of this discontinuity becomes evident at about the point in the stellar population where molecules can form in stellar photospheres, which supports a published model of molecular consciousness. A mechanistic explanation for Parenago's Discontinuity requires interaction between star fields and dense diffuse nebulae that might drag less massive stars faster than more massive stars. Reference to three major catalogs of deep-sky objects reveals that nebulae large enough to drag stars over a radius of ~260 light years are very rare. In 2016, the European Space Agency (ESA) released the first data set from the Gaia space observatory. It now appears clear that Parenago's Discontinuity is a non-local phenomenon. An unexpected and provocative aspect of the reduced Gaia observations is an indication that stars within >500 light years of the Sun apparently accelerate in the direction of their galactic revolution as they age. Other published supporting work includes a demonstration that certain binary stars are similar in many respects to biological organisms. A well-developed model of quantum consciousness supports the concept that neutron stars are conscious. It has been noted that spiral galaxies such as our Milky Way retain their shape after absorbing smaller satellite galaxies, which also supports panpsychism or self-organization at the highest levels. Another research team has reported that the apparent fractal arrangement of galaxy clusters and voids is also in congruence with some form of universe self-organization. The work of two separate researchers who are contemplating methods of communicating with stellar-level consciousness indicates that experimental astro-panpsychism may be possible as well as observational astro-panpsychism. It is becoming evident that panpsychism may be moving from the realm of metaphysics to the domain of observational astrophysics.

Keywords: Panpsychism, universe self-organization, Parenago's discontinuity, observational science.

[*]Correspondence: Prof. Gregory L. Matloff, Physics Dept, New York City College of Technology, CUNY, Brooklyn, NY, USA.
http://www.gregmatloff.com E-mail: GMatloff@citytech.cuny.edu

Introduction

In a previous article (Matloff, 2016), it was argued that aspects of panpsychism - The metaphysical doctrine that consciousness permeates the universe at all level of organization - could be tested in the astrophysical realm. As discussed in that article and elsewhere (Matloff, 2012, 2015, 2015a, 2017), the observational support for this thesis resulted from a consideration of stellar kinematics as part of a 2011 symposium at the London headquarters of the British Interplanetary Society to discuss the contributions of Olaf Stapledon, a British philosopher and visionary science-fiction author.

The aspect of Stapledon's work tested in this series of papers was his metaphysical doctrine that stars in a sense are conscious and that a portion of their motion is volitional. Jantsch (1980) has argued that stellar consciousness is to be expected in the context of universal self-organization. A neuronal basis of consciousness utilizing quantum tunneling , as argued by Walker (1970, 1999) and others is impossible for stars. The same is true for theories of consciousness based upon quantum entanglement among microtubules in a brain (Margolis, 2001, Hameroff and Penrose, 2014).

But there is a third possibility. It has been suggested that fluctuations in the quantum foam (also called the quantum vacuum) might be the origin of a proto-consciousness field that permeates the universe (Haisch, 2006). It has been known since 1948 that a significant fraction of the Van der Waal molecular bond is due to vacuum fluctuation pressure - the so-called Casimir Effect (Genz, 1999). The conjecture that consciousness arises from vacuum fluctuations is far from unreasonable - the Big Bang (certainly the most creative incident in the universe) - is thought to have originated from a stabilized vacuum fluctuation.

After selecting the possible conceptual model of how consciousness might enter the upper layers of a star, it was necessary to first determine at what point in the stellar population molecules might form. The next step was to see if any kinematics anomaly appears at or near this point. In other words, do stars with molecules in their upper layers (which might be conscious) move differently from stars too hot to have stable molecules.

As discussed below, observational spectroscopic research dating to the 1930's demonstrated that molecules begin to appear in the outer layers of stars a bit hotter than our Sun. A Soviet-era Russian astronomer, Pavel Parenago, demonstrated that at least for relatively near stars, stars with molecules revolve a bit faster around the center of our galaxy than their hotter sisters.

The Onset of Molecular Signatures in Stellar Spectra

A good place to begin a consideration of molecular signatures in stellar spectra is to discuss spectral classes, surface temperatures and color indices. Deep in the interior of mature main sequence stars, temperatures are uniformly very high due to the conversion of hydrogen to helium and energy. But the stellar surface temperature can vary widely, as presented in Table 1 [Drilling and Landolt (2000), Tokunaga, (2000)].

The hottest stars are of spectral class O, which are bright, blue, massive and comparatively short lived. Low mass, dim, red, long-lived M-class stars are coolest.

Table 1. Surface Temperatures and (B-V) Color Indices For Various Main Sequence Star Spectral Classes

Star Spectral Class	Surface Temperature (degrees Kelvin)	(B-V) Color Index
O5	> 36,000	-0.33
B0	31,500	-0.30
B5	15,400	-0.17
A0	9,480	-0.02
A5	8,160	0.15
F0	7,020	0.30
F5	6,530	0.44
G0	5,930	0.58
G5	~5,680	0.68
K0	5,240	0.81
K5	4,340	1.15
M0	3,800	1.40
M5	3,030	1.64

There are very few spectral lines in the spectra of very hot stars. Spectral lines corresponding to electronic transitions in molecules become more common as star surface temperature decreases.

The (B-V) color index of a star is essentially a comparison of a star's brightness in the blue and yellow ranges of the visible spectrum. It is a quantifying tool in the study of star spectral classes. If , for example, an astronomer determines the (B-V) color index of a particular main-sequence star to be 0.68, she can be reasonably sure that the spectral class of that star is G5. Note that hot stars tend to have negative (B-V) color indices and cooler stars have positive (B-V) color indices. Our Sun is usually classified as a G2 main sequence star. Its surface temperature is 5777 degrees

Kelvin and its (B-V) color index is 0.650 (Livingston, 2000). As discussed in Matloff (2016), observations of the onset of molecular signatures in stellar spectra are discussed by Russel (1934), Renze and Hynek (1937) and Swings and Struve (1932). This early work has been reviewed by Tsuji (1986).

Observations were conducted for more than two dozen stars. These references report that the spectral signature of simple CH and CN molecules begin to appear in stars cooler than spectral class F8 (photospheric temperatures < 6500 Kelvin degrees).

Parenago's Discontinuity

The next step was to search for information regarding a difference in stellar motions between stars hotter than spectral class F8 and those cooler than that class. As discussed in Matloff (2016) and elsewhere, the first indication that such a phenomenon exists was uncovered by the Soviet-era observational astronomer Pavel Parenago (1906-1960). Parenago's observations revealed that at least for stars relatively close to the Sun, cool stars revolve around the center of the Milky Way galaxy a bit faster than hot ones. According to Binney et al. (1997), the sharp difference in stellar velocities, which is referred to as Parenago's Discontinuity, occurs around (B-V) = 0.62, which corresponds to late F or early G stars.

In Matloff (2012, 2015, 2016, 2017), data have been plotted to demonstrate Parenago's Discontinuity for main sequence stars. This plot, which is included as Fig. 1, presents results discussed in Gillmore and Zelik (2000) and Binney et al. (1997).

In Fig. 1, the Binney et al (1997) data is from observations of more than 6,000 main sequence stars by the European Space Agency's Hipparcos space observatory. These stars are all within a sphere with a radius of about 260 light years, centered on the Sun. The ordinate presents variation in star orbital velocity around the center of the Milky Way galaxy.

Note the bulge in the Fig. 1 data points between (B-V) ≈ 0.55 and (B-V) ≈ 0.9. The possible significance of this feature was not recognized by this author in previous papers on this subject. More will be said about this intriguing feature in subsequent sections of this paper.

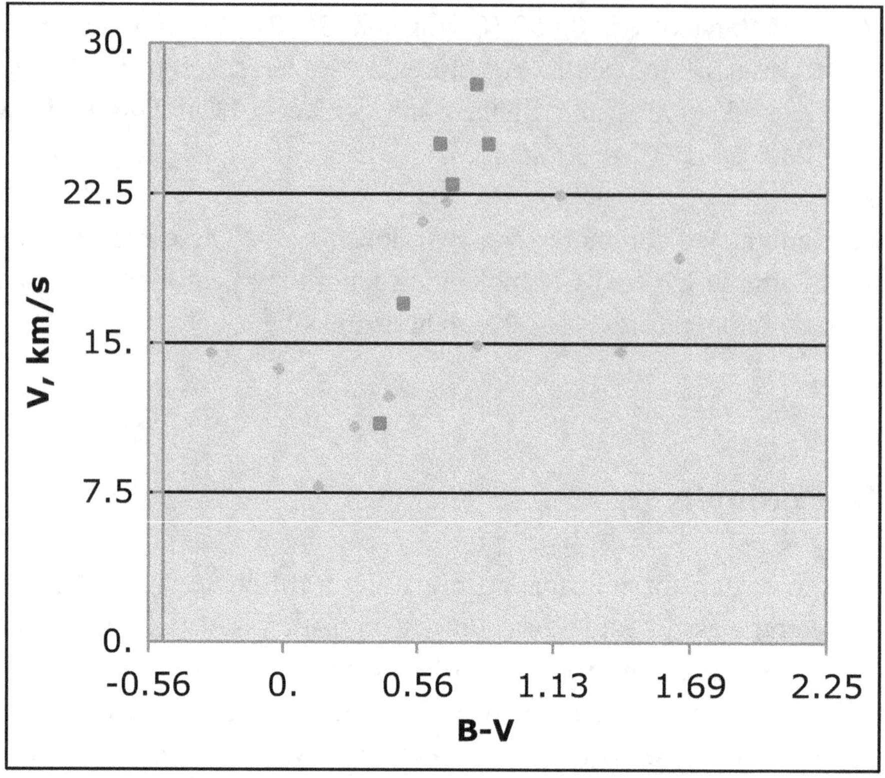

Fig. 1. Parenago's Discontinuity for Main Sequence Stars out to ~260 Light Years. Diamond Data Points are from Gilmore and Zelik (2000). Square Data Points are from Binney et al (1997).

Spiral Arms Density Waves: A Mechanistic Explanation for Parenago's Discontinuity - and Why it Likely Fails

In astrophysics, as in other disciplines, alternative explanations arise to explain observed natural phenomena. The leading mechanistic contender to explain Parenago's Discontinuity is Spiral Arms Density Waves (Binney, 2001 and DeSimone et al., 2004).

Reference to any photograph of a spiral galaxy such as our Milky Way reveals that there are dark patches in the star fields that compose the spiral arms. These are diffuse nebula, that are the nurseries in which infant stars form. The gas/dust density in these galactic clouds is typically more than 1000X greater than in the inter-cloud medium where our Sun resides.

It is not impossible that at various times, a diffuse nebula passes through our Sun's low-density galactic vicinity. According to the Spiral Arms hypothesis, during such an interaction low-mass, high (B-V) stars may be dragged along to a higher galactic-revolution velocity. by the more dense interstellar cloud. The fact that the velocity discontinuity occurs at or very near the point where the signature of stable molecules appears in stellar spectra is a coincidence.

There are two observational objections to the Spiral Arms Density Waves hypothesis. The first of these is a significant observational failure of the Spiral Arms hypothesis. For Spiral Arms to succeed as a hypothesis, there must be a color difference between stars near the leading and lagging edges of spiral arms in spiral galaxies similar to our Milky Way. Foyle et al. (2011) conducted an extensive spectroscopic study of twelve comparatively near spiral galaxies. No such color difference was observed.

The second objection derives from a consideration of the maximum size of catalogued diffuse nebulae. For Spiral Arms to succeed as an explanation of Parenago's Discontinuity from Hipparcos data, as presented in Fig. 1, at least some of these nebulae must have an effective diameter greater than 520 light years.

As discussed in Matloff (2016), three catalogs of deep-sky objects were studied to discover the sizes of the largest known diffuse nebulae in the Milky Way galaxy. These included a modern version of Charles Messier's 1784-vintage catalog of 102 objects (Jones, 1969) ,the much more comprehensive Herschel catalog with >2,500 objects (Mullaney and Tiron, 2011) and an on-line version of the very expansive *New General Catalog* (www.atlasoftheuniverse.com/nebulae.html).

These studies revealed that the median diameter of a diffuse nebula is less than 20 light years. Only 10% of the sample had diameters greater than 100 light years. The largest known diffuse nebula in our galaxy is the Eta Carinae circum-stellar nebula, which has a diameter greater than 400 light years.

The largest known diffuse nebula, 30 Doradus (the Tarantula Nebula) resides in an irregular galaxy (the Large Magellanic Cloud) which is a satellite of our Milky Way and is at a distance of about 200,000 light years. Burnham (1978) estimates the diameter of this object as 800 light years.

When Matloff (2016) was written, it therefore seemed unlikely that a diffuse nebular large enough to drag low mass stars in a sphere with a ~500 light year diameter exists. But because of the existence of one nebula large enough in a neighboring galaxy, it seemed impossible to rule out Spiral Arms.

To demonstrate whether Parenago's Discontinuity is a local or non-local phenomenon, it was necessary to obtain accurate distance and motion data for stars over a larger volume of space than was accomplished using Hipparcos. As discussed in Matloff (2015), the European Space Agency (ESA) launched the Gaia space observatory as a successor to Hipparcos in 2013. The goal of this mission is to determine accurate distances and motions for ~ 1 billion stars in the Milky Way galaxy. Some relevant data from the first Gaia data release are presented below.

Galactic Trajectory Alteration by Minded Stars

There is some discussion in Matloff (2012, 2015, and 2016) of how a minded star might alter its trajectory. One possibility is unipolar jets Emitted by infant stars, which have been observed (Namouni, 2007).

Applying a mass ejection rate consistent with that of young stars in the T Tauri phase (Gomez-de Castro et al. 2003), it was shown in Matloff (2016) that if 20% of a star's initial mass is ejected in a unidirectional jet at 100 km/s for the first billion years of a star's life, the star's velocity will be altered by about 20 kilometers per second.

A much more speculative possibility is also raised in Matloff (2016). If psychokinesis (PK) is scientifically verified, a very weak PK force exerted by a star over a billion-year time interval might be sufficient to alter the star's galactic-revolution velocity.

It was not thought in 2016 that data would soon be available relating to stellar volitional acceleration. But as discussed below, analysis of the first Gaia data release has revealed a very intriguing effect.

The balance of this paper is devoted to new material, in an effort to ascertain whether panpsychism in the celestial realm has emerged as a science. First the predictions in Matloff (2016) will be reviewed in light of the analysis of the first Gaia data release. Next the work of other researchers approaching this subject from multiple directions will be considered. Then, an approach to quantifying panpsychism will be discussed.

Astro-Panpsychism as a Science: Matloff (2016) Predictions Revisited

In the conclusion of Matloff (2016), it was noted that a scientific subject must be subject to experimental or observational validation / falsification. Four questions were listed regarding future observational results that could validate or falsify the concept. Surprisingly (and happily for this author), two of the questions have been answered. The answers of these apparently validate panpsychism. Consideration of a third in light of the first Gaia data release lends support to panpsychism and also leads to the possibility of an equally surprising alternative approach. To the author's knowledge, the fourth question has not been addressed by the scientific community.

Question 1: *Is Parenago's Discontinuity a Non-local, or Universal Phenomenon?*

In 2016, the European Space Agency released the first data release (DR1) from the Gaia space observatory (Gaia Collaboration, 2016 and Lindegren et al., 2016). Contained in Gaia DR1 are astrometric results for more than one billion stars in the Milky Way galaxy brighter than apparent visual magnitude 20.7. For about two million stars in this sample brighter than apparent visual magnitude ~11.5, positions, parallaxes and proper motions were measured with a precision equal to or greater than that of the Hipparcos data set.

This data set was applied by Vityazev et al. (2018) to investigate Parenago's Discontinuity over a much greater volume of space and for many more stars than was done using Hipparcos data (Binney et al, 1997) and Gilmore and Zelik (2000). The Vityazev et al. (2018) sample contains 1,260, 071 main sequence stars and 534,387 red giants. Because of the high luminosity of giant stars, many of these were quite distant. As a star's distance increases, the accuracy of its distance estimate decreases. It was not possible in this study to confirm the assertion of Branham (2011), which is discussed in Matloff (2016), that Parenago's Discontinuity exists for giant stars as well as main sequence stars. Some aspects of the Vityazev et al. (2018) main sequence star sample are presented in Table 2.

Table 2. Vityazev et al.(2018) Sample of Gaia DR1 Main Sequence Stars

Spectral Type	O	B	A	F	G	K	M
Mean (B-V)	-0.37	-0.06	0.19	0.46	0.68	0.96	1.49
Number	91	9165	189560	601577	425150	33843	625
Minimum Estimated Average Distance (kpc)	0.78	0.59	0.59	0.45	0.33	0.14	0.03
Largest Estimated Average Distance (kpc)	2.54	1.15	1.11	0.64	0.40	0.15	0.03
Average Star Age (X 10^9 years)	-	-	-	3.5	8.0	7.7	7.0

Note: 1 kpc = i kiloparsec= 1,000 parsecs or 3260 light years

The most luminous (and rarest) stars in Table 2 are O class. Because of their brightness, they tend to be at a great distance from our solar system. But the spread in average distance estimates for these stars is enormous - 2543 to 8280 light years. Less luminous B and A stars still emit a lot more electromagnetic energy than our Sun and the range in average distance estimates for these stars are less extreme. There are many more B and A stars in the Gaia DR1 main sequence sample than O stars.

Most stars in this population are F, G, and K stars. The spread in average distance estimates for these stars is a lot less than for hotter stars. Average F stars are within 1467 - 2086 light years from the Sun. Average K stars are within about 500 light years of the Sun. Solar-type G stars in this sample are at an average distance of 1085 - 1304 light years from our solar system.

Although the most numerous stellar spectral class is M, these red dwarfs are very dim. Because of the magnitude limitation of Gaia's instruments, there are few M stars in the sample. On average, M stars in the sample are within 100 light years of the Sun.

It is not surprising that the average age of F stars in this sample is 3.5 billion years, since stars in this spectral class do not live as long as our G-class Sun, which is about half-way through its projected 10-billion year life span. But it is very intriguing that most G and K main sequence stars in the Gaia DR1 sample are billions of years older than our Sun.

If intelligent, technological life has evolved and survived on planets orbiting G or K stars in our galactic vicinity, they are likely billions of years in advance of terrestrial civilization. It is humbling to realize that if we encounter extraterrestrials, they will likely be so far in advance of us that we might not even recognize them.

Figure 2 presents galactic revolution velocities for main sequence stars versus (B-V) color index from Fig. 1B of Vityazev et al. (2018). All or most of these 1,260, 071 stars reside in a sphere centered on the Sun with a diameter > 1,000 light years.

Note in Fig. 2 that the bulge in the Fig. 1 data points between (B-V) ≈ 0.55 and (B-V) ≈ 0.9 is clearly present, which indicates that this is a real feature. This interesting feature, which is likely real because it is in both data sets, is discussed in greater detail below.

The data point on the extreme left of the graph, which corresponds to O spectral class stars with an average (B-V) color index of about -0.4 do not likely revolve as fast as indicated. Reference to Fig. 1B of Vityazev (et al. (2016) indicates that the error bars in galactic revolution velocity V are enormous, likely because of the rarity of O stars in the galaxy and the large errors in distance estimates for this class of stars.

ISSN: 2153-8212   Journal of Consciousness Exploration & Research   www.JCER.com
                         Published by QuantumDream, Inc.

Note that Parenago's Discontinuity occurs in this figure at about the same (B-V) value where it occurs in the Fig. 1 data from Hipparcus and *Allen's Astrophysical Quantities* (Binney et al, 1997) and (Gilmore and Zelik ,2000).

Because Parenago's Discontinuity appears in the Fig. 2 data set, for stars > 500 light years from the Sun, it seems very unlikely that Spiral Arms Density Waves is a viable explanation. Parenago's Discontinuity is clearly a non-local phenomenon.

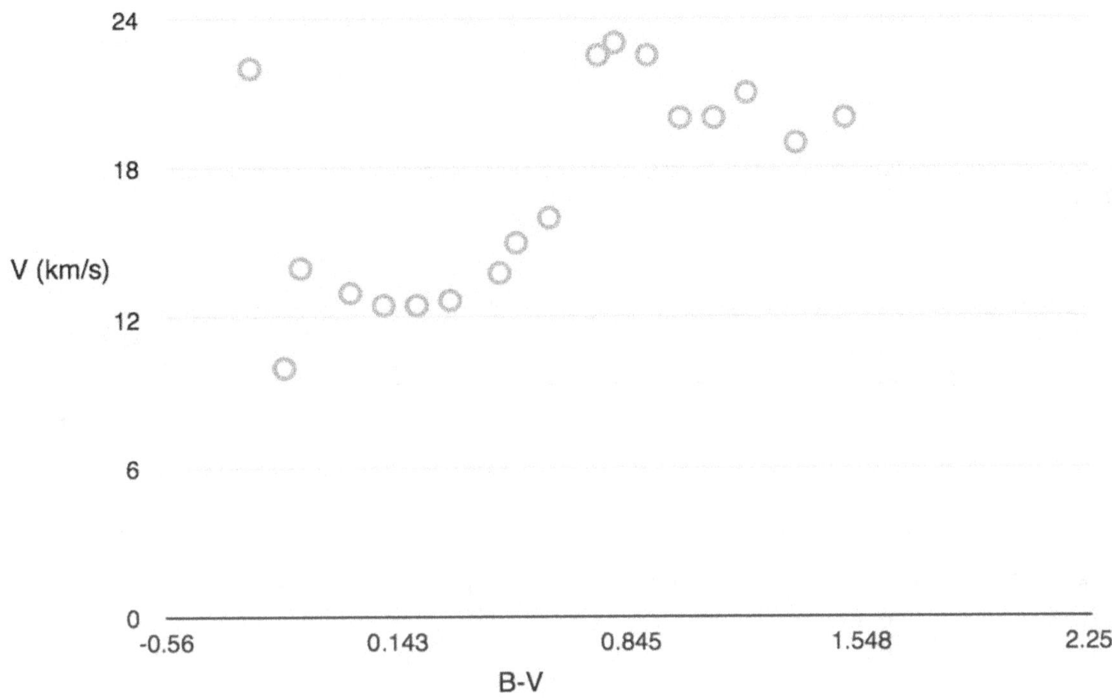

Fig. 2. Relative Star Motion in the Direction of the Sun's Galactic Revolution (V) vs. (B-V) color Index from Gaia DR1 Data (Vityazev, et al. 2018).

When the data sets in Figs 1 and 2 are plotted together, Parenago's Discontinuity is clearly revealed. It is also evident that data points from both data sets line-up rather well. This is presented in Fig. 3.

It shouldn't be concluded that these results entirely rule out a mechanistic explanation for Parenago's Discontinuity. But constructing such a hypothesis will require a great deal of ingenuity. As is the case with Spiral Arms, such a mechanistic hypothesis may not survive future efforts to demonstrate that Parenago's Discontinuity is not merely a non-local phenomenon but a galactic phenomenon. It is also hoped that future researchers attempt to observe Parenago's Discontinuity in the motions of stars in galaxies external to our Milky Way.

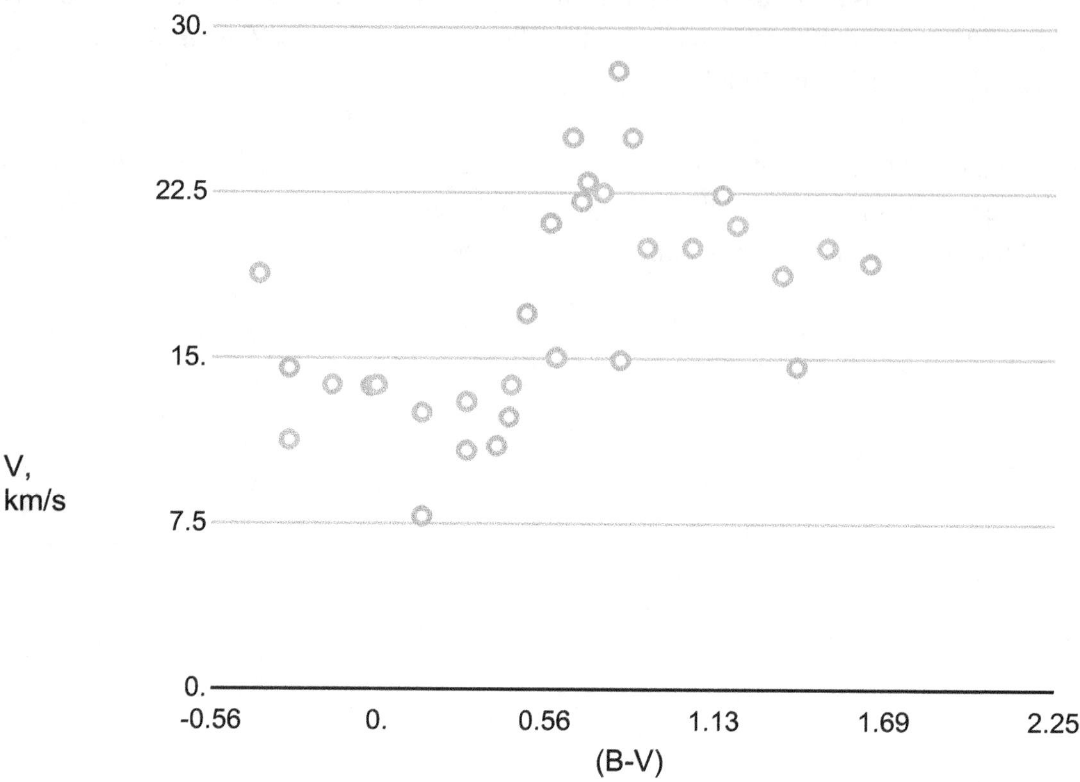

Fig. 3. Combined Plot of Fig. 1 Data (Blue) from Hipparchus and Allen's Astrophysical Quantities with Fig. 2 Gaia DR1 Data (Green).

<u>Question 2</u>: *What is revealed by Further Study of Unipolar Stellar Jets?*

When I posed this question during the preparation of Matloff (2016), I did not expect results anytime soon. As discussed above, the most likely acceleration process for a minded star seemed at the time to be a unipolar jet of material that has been observed to be emitted by infant stars. Might future observation of these jets indicate that they tend to be in a direction opposite to the star's direction of galactic revolution? Might the intensity of the jet vary with an infant star's distance from the galactic center?

But a major result of the very first study of Parenago's Discontinuity using Gaia DR1 data (Vityazev, et al. 2018) is very relevant to the study of minded star acceleration and perhaps as well to considerations of the role of technologically advanced conscious life in the universe.

As well as validating the existence of Parenago's Discontinuity over a much greater volume of space than previous studies could achieve, Vityazev, et al. (2018) were able to investigate the curious bump in Figs. 1-3 between (B-V) ≈ 0.55 and (B-V) ≈ 0.9. They were able to do this because of the large Gaia DR1 star sample and the age estimates for stars in this sample that are presented in Table 2 above.

To my great surprise, the analysis of Vityazev et al. (2018) indicates that mature stars accelerate along the direction of their galactic revolution (and only in this direction) as they age! This is not a tremendous acceleration - it amounts to an increase in galactic revolution velocity of about 1 kilometer per second in every billion years or about 3×10^{-14} meters per second squared. The acceleration of gravity at Earth's surface is about 3×10^{14} times greater than this stellar acceleration!

After the publication of Vityazev et al. (2018), veteran science-journalist Paul Glister decided to devote the April 26, 2019 issue of his well-researched astrophysics/astronautics blog www.centauri-dreams.org to the subject of Parenago's Discontinuity.

My two previous blog contributions on this subject had been organized as a written statement by me followed by a discussion with blog respondents. In this case, Glister decided to conduct the blog entry, which was entitled "Probing Parenago: A Dialogue on Stellar Discontinuity", as a dialogue between me and Alex Tolley (a frequent blog contributor) followed by a discussion with various respondents.

When we debated the apparent acceleration of mature stars, I initially considered mechanisms that the mature, minded star might use to increase its velocity of galactic revolution. These included (but were not limited to) unidirectional electromagnetic radiant output and a directed solar wind. I initially ruled out Coronal Mass Ejections (CMEs), although Kelvin Long (2020) has informed me recently that the velocity of some CME events is higher than 3,000 kilometers per second.

It was a surprise when Tolley, echoed by several blog respondents, suggested an alternative to minded stars to explain this apparent acceleration of mature star. As reviewed in Table 2, most main sequence stars in the Gaia DR1 sample are billions of years older than our Sun. This is sufficient time for even a few very-long-lived extraterrestrial civilizations capable of conducting interstellar colonization missions like those we can envision today to have occupied most or all planetary systems in the galaxy. This concept, which I have considered further in Matloff (2019) leads to the possibility that we live in an occupied and engineered galaxy.

Assuming that further observational research confirms this stellar acceleration, astrophysicists might have to choose between the possibility of minded stars or ET-directed stars, if some mechanistic explanation for this stellar acceleration does not succeed. Either possibility would be humbling to those humans who like to picture themselves as lords of creation.

Question 3: *Does Further Research Support the Conclusion that Molecules are Found in Stars cooler than F8?*

As discussed above and in Matloff (2016), the sample of stars studied in the 1930's to determine the spectral class of stars in which molecular spectral signatures first appear is very small. It is hoped that the much larger library of stellar spectra available today can be accessed to check the conclusion that spectra of stars cooler than spectral class F8 contain signatures of simple molecules such as CH and CN. To my knowledge, results of such studies have not yet been published.

Question 4: *Might There be Observational Signs of Panpsychism or Self-organization at Higher Cosmic Levels?*

In Matloff (2016), I refer to the conjecture of Freeman Dyson that the Mind is present at many levels in the universe (Dyson, 1988). One possible example of self-organization at the galactic level is the observation of Ari Maller (2007) that spiral galaxies such as our Milky Way tend to retain their spiral shapes even after repeatedly absorbing smaller galaxies.

A Romanian philosopher and physicist have collaborated on a study of structures in the universe that are larger than galaxies including galaxy clusters, filaments, and voids. According to their conclusions, the fractal nature of these constructs points towards self-organization on a grand scale (Murdzek and Iftimie, 2008). More recently, Renyue Cen (2014) of Princeton University has found evidence of self-organization in galaxy formation.

An end-of-life stage for stars considerably more massive than our Sun is the compact neutron star. According to D. K. Berry et al. (2016), there is a striking similarity between biological cellular structures and simulated structures in neutron stars. Penrose and Hameroff (2001) speculate that according to their theoretical studies of quantum consciousness, neutron stars may be conscious.

Interacting binary stars are stellar couples so close than one star seems to live off its companion. Clement Vidal (2014, 2016) has conducted a comprehensive analysis of these systems. He concludes that these interacting stellar pairs share many of the properties of biological systems.

Evolutionary biologist Stuart Kauffman (1993) and others have reported evidence for self-organization at much smaller biological levels. It would be nice if a quantitative mechanism could be devoted to consider self-organization at levels from the molecule to the universe.

ISSN: 2153-8212   Journal of Consciousness Exploration & Research   www.JCER.com
Published by QuantumDream, Inc.

One person who is addressing this issue is Giulio Tononi (2012a, 2012b). His approach, Integrated Information Theory, treats consciousness as an intrinsic property of any physical system. The consciousness level of any system depends upon the number of interconnections allowing the integration of information. The brain of a human or higher animal has billions of neurons with a vast number of interconnection possibilities. The level of consciousness of such an organism is enormously greater than that of a molecule, which has a small number of interconnection possibilities.

My earlier work on anomalous stellar motions Matloff (2012, 2015, 2015a, 2016, 2017) assumed that stable molecules in or above a star's photosphere have sufficient interconnection possibilities to enable a type of flocking behavior. But there are a lot of molecules in the outer layers of a Sun-like star. So it is certainly not impossible that the consciousness level of such a star is enormous.

Gregory Benford, a physicist affiliated with The University of California at Irvine, has investigated the possibility of god-like stars in a co-authored science-fiction novel (Benford and Eklund, 1977). But could we communicate with a minded star?

One researcher considering modes of establishing a communication link with the Sun is the British biologist Rupert Sheldrake. After receiving his Ph.D. from Cambridge University, Sheldrake worked as a plant physiologist in India. During his stay in India, Sheldrake became acquainted with aspects of Hinduism including Sun mantras and salutations. He speculates whether we could apply such techniques to provoke a solar response, such as a selected sunspot pattern (Sheldrake, 2018).

Van de Bogart (2017, has suggested an alternative possible human-stellar communication mode. Because sound - especially those rhythmic patterns we call songs - are so significant among terrestrial life forms, perhaps stars communicate using subtle variations in their electromagnetic output. Perhaps astrophysicists working in helioseismology could uncover such patterns.

Conclusions: Is Panpsychism Metaphysics or an Observational Science?

We can now discuss the central theme of this paper: Has Panpsychism emerged from metaphysics to enter the realm of science? A number of philosophers and scientists have argued the same point. A recent example is a publication of Philip Goff (2019), a philosopher at Durham University in the UK.

There are several criteria that a discipline must satisfy to be considered scientific. First, can cogent hypotheses be developed to explain physical phenomenon. Second, can these hypotheses be tested by experiment or observation and perhaps evolve into an accepted theory. Next, it should be possible to successfully predict the results of future observations or experiments based upon the hypotheses developed. As the discipline matures, quantitive processes should be developed and applied. Many people should be attracted to the new field and begin to contribute to it. Unexpected significant results should emerge from these studies. Finally, the new scientific discipline should begin to influence the wider culture of our civilization. Based upon the discussion above, we next see how Astro-Panpsychism currently succeeds in satisfying these criteria.

Criteria 1: Development of a Hypothesis related to Astro-Panpsychism

As first described in Matloff (2012) and further discussed in my subsequent work, constructing the hypothesis that a portion of stellar motion is volitional was readily accomplished. A literature search revealed Parenago's Discontinuity, a phenomenon that supported this hypothesis. Further work revealed a molecule-based theory of consciousness and demonstrated that the velocity discontinuity in stellar revolution velocity around the galaxy's center is at or near the point where the signatures of simple molecules appear in stellar spectra.

Criteria 2: Can this hypothesis be tested by further experiment or observation?

An alternative, mechanistic concept to explain Parenago's Discontinuity (Spiral Arms Density Waves) requires large diffuse nebulae interacting with a star field to drag relatively low-mass stars (such as the Sun) more rapidly than more massive stars. At the time that the above hypothesis was generated, Parenago's Discontinuity was observationally evident for main sequence stars within about 260 light years of the Sun. A search of three catalogs of deep sky objects revealed only one diffuse nebula (30 Doradus) large enough to satisfy the requirements of Spiral Arms Density Waves. it was suggested that observations of star positions and motions using the European space Agency's Gaia space observatory could reveal whether 30 Doradus,
the largest diffuse nebula known among galaxy neighbors to our Milky Way, would be large enough to explain Parenago's Discontinuity if Gaia results indicated that this phenomenon is non-local.

Criteria 3: It should be possible to successfully predict the result of future observations based upon the hypothesis in Criteria 1

Further observations using the first data release from the Gaia space observatory revealed that Parenago's Discontinuity is observed for main sequence stars over a much greater radius than that of 30 Doradus. This result apparently falsifies Spiral Arms Density Waves, the competing mechanistic hypothesis to the one discussed in Criteria 1.

<u>Criteria 4</u>: Quantitative processes should be developed and applied to any new scientific discipline

The "PHI" approach (Tononi, 2012a, 2012b) is an excellent first step in the quantification of Panpsychism.

<u>Criteria 5</u>: Many people should be attracted to the new discipline and contribute to it

See the cited works by Maller, (2007), Murdzek and Iftimie (2008), Sheldrake (2018, Van de Bogart, (2017), and Vidal (2014, 2016).

<u>Criteria 6</u>: Unexpected significant results should emerge from these studies

See the cited work by Vityazev et al. (2018). As well as the expected result that Parenago's Discontinuity is a non-local phenomenon, this paper presents totally surprising and possibly very significant data demonstrating that main-sequence stars accelerate along their galactic trajectories as they age. This remarkable result, which points to either an engineered or a conscious galaxy, will hopefully be replicated by future observations.

<u>Criteria 7</u>: The new scientific discipline should begin to influence the wider culture of our civilization

The concept of Astro-Panpsychism has long-since entered the genre of science fiction in the cited novel Benford and Eklund (1977). Two more recent science-fiction novels that further explore this concept, as well as presenting a case for stellar acceleration are authored by Gregory Benford and Larry Niven. These are *Bowl of Heaven*, (New York, Tor, 2012) and *Shipstar* (New York, Tor, 2014).

In an equally significant recent publication, Panpsychism is a plot element in a successful techno-thriller. This novel, authored by Julian Gough, is entitled *Connect* (New York: Doubleday, 2018).

Elements of Panpsychism have also appeared in other works of prose and poetry. See for example Alison Hawthorne Deming's *The Edges of the Civilized World: A Journey in Nature and Culture* (New York: Picador, 1999).

Panpsychism clearly satisfies all of the above criteria. So it is hard to argue that it is not emerging as a scientific discipline. It is clearly possible though, that it might ultimately be superseded by some other metaphysical principle. For example,Deepak Chopra and Menas Kafatos argue in *You Are The Universe* (New York: Harmony, 2017) that it will ultimately be replaced by Idealism, in which consciousness is not only present at all universal levels but is dominant over matter and energy.

Acknowledgements:

I have never been as far from my comfort zone as an applied space scientist specializing in space propulsion technologies as I have in this and previous considerations of Astro-Panpsychism. There were times when the challenges of attempting to birth this new scientific discipline were most daunting. Without the encouragement and inspiration of my wife and partner, artist C Bangs, this task might have been impossible.

The assistance and enthusiasm of Kelvin Long, the physicist who directs The Interstellar Research Centre in Gloustershire, UK is also acknowledged. Kelvin, who at the time was Editor of *The Journal of the British Interplanetary Society (JBIS)*, graciously delivered my paper that became Matloff (2012) at a 2011 symposium at BIS headquarters in London devoted to the work of Olaf Stapledon. In 2019 and early 2020, Kelvin devoted a great deal of effort to the preparation for a workshop devoted to the topic of consciousness in the universe. Planned for June 2020 and now indefinitely postponed because of the pandemic, correspondence with other planned participants was instrumental in planning this paper.

During June 2019, C and I visited Rupert Sheldrake in his London home. Discussions with him regarding techniques for communicating with minded stars and the possible implications of such communication have been inspiring.

We were also visited in New York by Julian Gough, the author of *Connect*. It is hoped that this techno-thriller and other popular publications can successfully communicate the essence of Panpsychism to a mass audience.

Finally, I would like to acknowledge the assistance of my friend and late mentor Evan Harris Walker. Although I initially became acquainted with Harris during a calibration on an advanced space-propulsion concept, my early discussion of quantum consciousness theories owes a great deal to his pioneering work in that field. Two of his relevant publications are cited here.

Received June 29, 2020; Accepted July 25, 2020

References

Benford, G. and Eklund, G. (1977): *If the Stars are Gods*, New York, Berkley-Putnam.

Berry, D. K., Caplan, M. E., Horowitz, C. J., Huber, G., and Schneider, A. S. (2016.): "Parking Garage Structures in Nuclear Astrophysics and Cellular Biophysics. " in *Physical Review C*, vol. 94, 055801,

Binney, J. J., Dehnen, W., Houk, N., Murray, C. A., Preston, M. J. (1997): "Kinematics of Main Sequence Stars from Hipparcos Data", in *Proceedings ESA Symposium Hipparcos Venice 97, ESA SP-402, Venice, Italy, May 13-16, 1997*, 473-477. Paris, European Space Agency.

Binney, J. J. (2001), "Secular Evolution of the Galactic Disk", in *Galaxy Disks and Disk Galaxies*, eds. F. Bertoli and G. Coyne, ASP Conference Series vol. 230. San Francisco, Astronomical Society of the Pacific.

Branham, R. L.(2011), "The Kinematics and Velocity Distribution of the GIII Stars: *Revisita*

Mexicana de Astronomia y Astrofisica, vol. 47, 197-209.

Burnham, Jr. (1978): *Burnham's Celestial Handbook*, New York: Dover.

Cen, R. (2016), "Temporal Self-Organization in Galaxy Formation." In *Asrophysical Journal Letters*, Vol. 785:L21.

DeSimone, R. S., Wu, X., Tremaine S. (2004): "The Stellar Velocity Distribution in the Stellar Neighborhood" In *Monthly Notices of the Royal Astronomical Society*, vol. 350, 627-643.

Drilling, J. H. & Landolt, A. U. (2000). "Normal Stars", Chap. 15 in *Allen's Astrophysical Quantities, 4th Edition*, ed. A. N. Cox, New York: Springer-Verlag.

Dyson, F. (1988): *Infinite in all Directions*. New York: Harper & Row.

Foyle, K., Rix, H.-W., Dobbs, C., Leroy, A., Walter, F. (2011): "Observational Evidence Against Long-Lived Spiral Arms in Galaxies". In *Astrophysical Journal*, vol. 735, Issue 2, Article ID = 101 (2011).

Gaia Collaboration, A., et al. (2016): "Gaia Data Release 1: Summary of the Astrometric, Photometric, and Survey Properties", in *Astronomy and Astrophysics,* Manuscript no. aa29512-16 (Sept. 19, 2016).

Genz, H. (1999). Nothingness - The Science of Empty Space. Reading, MA: Perseus Books.

Gilmore, G. & Zelik, M. (2000): "Star Populations and the Solar Neighborhood", Chap. 19 in *Allen's Astrophysical Quantities, 4th Edition*, ed. A. N. Cox, New York: Springer-Verlag.

Goff, P. (2019): Galileo's Error: Foundations for a New Science of Consciousness, New York: Penguin/ Random-House.

Gomez de-Castro, A. I., Verdugo, E., Ferro-Fontan, C. (2003): "Wind Jet Formation in T Tauri Stars: Theory vs. UV Observation." In *Proceedings of 12th Workshop on Cool Stars, Stellar Systems and the Sun, July 30-August 3, 2001,* Boulder CO: University of Colorado Press.

Haisch, B. (2006). *The God Theory: Universes, Zero-Point Fields and What's Behind it All.* San Francisco: Weiser Books.

Hameroff, S. & Penrose, R. (2014). "Consciousness in the Universe: A Review of the Orch OR Theory". In *Physics of Life Reviews,* vol. 11, 39-78.

Jantsch. E. (1980): *The Self-Organizing Universe: Scientific and Hunan Implications for the Emerging Paradigm of Evolution.* New York: Pergamon.

Jones, K. G. (1969): *Messier's Nebulae and Star Clusters.* New York: American Elsevier Publishing Company.

Kauffman, S. (1993): *Origins of Order: Self-Organization and Selection in Evolution.* New York: Oxford University Press.

Lindgren, L., Lammers, U., Bastian, U., et al. (2016): "Gaia Data Release 1: Astrometry- One Billion Positions, T0o Million Proper Motions and Parallaxes". In Astronomy & Astrophysics, Manuscript No. aa201628714 (Sept. 15, 2016), arXiv:1609.04303v1 [astro-ph.GA1] 14 Sep 2016.

Livingston, W. C. (2000),:Chap. 14 in *Allen's Astrophysical Quantities, 4th Edition*, ed. A. N. Cox, New York: Springer-Verlag.

Long, K. F. (2020). "Comments on the Volitional Motion of Stars via Self-Thrusting of Stellar Winds and Coronal Mass Ejections." Paper in Draft.

Maller, A, (2007): "Halo Mergers, Galaxy Mergers, and Why Hubble Type Depends on Mass". Presented at *Formation and Evolution of Galaxy Disks*, Vatican Observatory, Rome, October 1-5, 2007.

Margolis, L. (2001): "The Conscious Cell". In *Cajal and Consciousness, Annals of the New York Academy of Sciences*, ed. P. J. Marjian, vol. 929, pp. 55-70.

Matloff, G. L. (2012): "Invited Commentary: Olaf Stapledon and Conscious Stars". In *JBIS*, vol. 65, 5-6.

Matloff, G. (2015): *Starlight, Starbright: Are Stars Conscious?*. Norwich, UK: Curtis Press.

Matloff, G. L. (2015a): "The Nonlocality of Parenago's Discontinuity and Universal Self-Organization". In *Axiom: The Journal of the Initiative for Interstellar Studies*, vol. 1, 14-19. Also presented at both International Academy of Astronautics (IAA) Symposium on the Future of Space Exploration: Towards New Global Programmes, Turin, Italy, July 7-9, 2015.

Matloff, G. L. (2016): "Can Panpsychism Become an Observational Science" In *Journal of Consciousness Exploration and Research*, vol. 7, 524-543.

Matloff, G. L. (March, 2017): "Stellar Consciousness: Can Panpsychism Emerge as an Observational Science", In *EdgeScience*, Number 29, 15-20.

Matloff, G. L., (2019): "Is the Kuiper Belt Inhabited?", *JBIS,* vol. 72, 382-385.

Mullaney J. & Tirion, W. (2011): The Cambridge Atlas of Herschel Objects. Cambridge UK: Cambridge University Press.

Murdzek, R. and Iftimie, O. (2008): "The Self-Organizing Universe", in *Romanian Journal of Physics*, vol. 53, 601-606.

Namouni, F. (2007): "On the Flaring of Jet-Sustaining Accretion Disks". In *Astrophysical Journal,* vol. 659, 1505-1510.

Penrose, R, and Hameroff, S. (2001): "Consciousness in the Universe: Neuroscience, Quantum Space-Time Geometry and the Orch OR Theory.", in *Journal of Cosmology*, vol. 14 (http:/ journalofcosmology.com/Consciousness160.html

Rense, W. A. & Hynek, J. A. (1937): "Photometry of the G Band in Representative Stellar Spectra". In *Astrophysical Journal,* vol. 86, 460-469

Russell, H. R. (1934): "Molecules in the Sun and Stars". In *Astrophysical Journal,* vol. 79, 317- 342.

Sheldrake, R. (2018): "Is the Sun Conscious?". Presented at Electric Universe Society meeting, Bath UK ((July 7, 2018). This lecture can be accessed using YouTube.

Swings, P. & Struve (1932): The Bands of CH and CN in Stellar Spectra" In *Physical Review*, vol. 39, 142-150.

Tokunaga, A. T. (2000): "Infrared Astronomy", Chap. 7 in *Allen's Astrophysical Quantities, 4th Edition*, ed. A. N. Cox, New York: Springer-Verlag.

Tononi, G. (2012a): "Integrated Information Theory of Consciousness". In *Archives Italiennes de Biologie*, vol. 150, 290-326.

Tononi, G. 2012b: *PHI: A Voyage from the Brain to the Soul*. New York: Pantheon.

Tsuji, T. (1986): "Molecules in Stars". In *Annual Review of Astronomy and Astrophysics,* vol. 24, 197-234.

Van de Bogart, W. G. (2017): "Orchestration of Consciousness: Consciousness and Electronic Music Applied to Xenolinguistics". Presented at Consciousness Reframed Conference, Beijing, China, November 25-27, 2017.

Vidal, C. 2014. *The Beginning and the End: The Meaning of Life in a Cosmological Perspective.* New York: Springer.

Vidal, C. 2016. Stellivore Extraterrestrials: Binary Stars as Living Systems, in *Acta Astronautica,* vol. 128, 251-256.

Vityazev, V. V., Popov, A. V., Tsvetkov, A. S., Petrov, S. D., Trofimov, D. A., and Kiyaev, V. I. (2018): "New Features of Parenago's Discontinuity from Gaia DR1 Data". In *Astronomy Letters*, vol. 44, 629-644.

Walker, E. H., (1970): "The Nature of Consciousness." In *Mathematical Biosciences*, vol. 7, 131-178.

Walker, E. H. (1999): *The Physics of Consciousness*. Cambridge MA: Perseus Books.

Research Essay

Quantum Theory & Consciousness

Bradley Y. Bartholomew[*]

Abstract

A recent research paper entitled "Generating mechanical and optomechanical entanglement via pulsed interaction and measurement" is paving the way towards demonstrating that even macroscopic harmonic oscillators can display this 'spooky' phenomenon in quantum mechanics of being 'entangled', that is to say information is passing between these harmonic oscillators faster than the speed of light. The neurons in the brain are in fact harmonic oscillators and it is argued that the 'binding problem' in cognitive science is solved if it turns out that all the neurons in the neural network are in fact entangled harmonic oscillators which means that information from different neurons or sets of neurons from diverse regions of the brain can integrate simultaneously to generate in us a unified or holistic consciousness. The philosophical implications of some of the 'weird' aspects of quantum mechanics are discussed.

Keywords: Niels Bohr, Copenhagen Interpretation, wave function collapse, entanglement, brainwaves.

The phenomenon of entanglement in quantum mechanics is well known even by the general public, and is poorly understood by non-experts and experts alike. Even Einstein was baffled by it and famously referred to it as 'spooky'. Basically quantum particles seem to defy our fundamental conceptions that we are living in a real physical world where there are 'causes' and 'effects' which occur after the lapse of a certain amount of time and in a defined space in the universe. Entangled particles defy the classical laws of physics in as much as they appear to act on each other instantaneously over distances that are clearly impossible in the real world. Originally it was thought to be only particles at the quantum level that displayed this peculiar characteristic which begs the question whether at the quantum level time and space actually exist at all. It has now been proved beyond doubt that many subatomic particles and indeed even atoms and larger molecules can display this property of entanglement and there is very active research as to just how big objects have to be before they cease to display this 'spooky action at a distance'.

How to entangle macroscopic objects using particles of light

A recent research paper has proposed a new way of creating entanglement between mechanical motion and an optical field, and also between two mechanical oscillators, using short pulses of light.[1] The authors say: "Making an entangled state of something bigger allows us to test

---

[*] Correspondence: Bradley Y. Bartholomew, Independent Researcher, France. Email: brad.bartholomew2@gmail.com

quantum physics on a macroscopic scale and paves the way for the development of powerful new quantum technologies, such as sensing and quantum networking. In this work, we propose a new technique to create such quantum states by using quantum optomechanics. In our scheme, light bounces off two tiny mechanical oscillators causing them to move, creating an entangled state in their motion."[2]

According to the authors: "It is a very exciting time in quantum optomechanics at present, as early signatures of quantum motion of mechanical oscillators are now being observed. In this work, we theoretically proposed and analysed a powerful new approach to create entanglement between mechanical motion and an optical field, and also between two mechanical oscillators, using short pulses of light."[2]

This research is the first theoretical proposal for preparing and verifying mechanical and optomechanical entanglement in the emerging field of pulsed optomechanics. This highlights a route to observe quantum entanglement in a regime where it hasn't been seen before. More specifically, there are two main regimes in quantum optomechanics. One is the "resolved-sideband regime", where light circulates inside an optical cavity for a timescale that is long compared to the mechanical period. The other is the "unresolved-sideband regime", which allows for rapid pulsed interactions over a timescale much shorter than the mechanical period. Optomechanical entanglement has not yet been observed in this latter regime and this research makes key steps in this direction. The authors report that one of the most important results from their work is that in the pulsed regime creating such entanglement looks experimentally accessible, and is robust to effects that often hinder its observation. And most importantly further advances in this research will address very fundamental questions, such as "what are the limits of quantum theory?" To these ends, a key current goal of research is to create and observe entanglement on larger and larger mass scales.[2]

Studying the various forms of optomechanical entanglement is an active current area of research and there have been several theoretical, and more recently experimental, studies exploring this avenue. While we do not aim to provide a thorough review here, we briefly describe and contrast some key studies of both optical–mechanical and mechanical–mechanical entanglement.

Focussing first on theoretical proposals, a notable early direction on optomechanical entanglement was to use photon–phonon transfer operations, where non-classical and entangled states of light are mapped onto the motion of mechanical oscillators via light–matter beamsplitter interactions.

Interactions of this type require operation in the resolved-sideband regime of optomechanics, which is realized if the mechanical frequency far exceeds the cavity decay rate. Following these proposals, detailed studies into the generation and detection of Gaussian optical–mechanical entanglement have been carried out, which focus on the linearized dynamics around the steady state. In the steady state, optomechanical systems are subject to certain stability conditions, which preclude parts of parameter space, particularly for optical drive on the blue sideband. Long-pulsed optical drives operating in the resolved-sideband regime were suggested as a means to access this part of parameter space, and the effect of high thermal occupation, multiple interactions, pulse shaping, and different optical detunings have also been studied for such long optical pulses.

In addition to optical–mechanical entanglement, developing methods to establish entanglement between a pair of mechanical oscillators is also of key interest. In particular, the steady-state linearized dynamics of two mechanical oscillators interacting with the same entangling optical field have been studied to investigate Gaussian entanglement. Alternative experimental configurations have also been proposed, such as the suspension of two mechanical membranes within the same optical cavity, and the injection of squeezed light into both a double cavity system and a ring cavity design. The generation of mechanical entanglement via reservoir engineering of a single cavity mode by a multi-tone drive has also been suggested. Recently, a more comprehensive analysis of steady-state Gaussian entanglement has been carried out, which investigates optical–mechanical, mechanical–mechanical, and tripartite light-mechanics entanglement. Importantly, this analysis goes beyond the usual resolved-sideband regime and does not employ the rotating-wave or adiabatic approximations typically used in the literature.

Swapping optical–mechanical to mechanical entanglement by long-pulsed interactions, optical interferometry, and measurements provides another exciting route for generating entangled states of mechanical oscillators. In this setting, linear interactions and measurements on the optical field have been discussed as a way to generate Gaussian continuous-variable entanglement. Furthermore, complementary approaches that implement photon-counting measurements to generate and witness non-Gaussian mechanical entanglement have also been considered. Additionally, recent theoretical work, which proposes entanglement swapping between two optical and two mechanical modes offers a promising avenue to use mechanical entanglement to investigate spatially dependent decoherence in massive systems.

In recent years, field–mechanics and mechanics–mechanics entanglement experiments have been performed, demonstrating the interest in, and feasibility of, generating optomechanical entanglement. Notably, long-pulsed optomechanical interactions in the resolved-sideband regime have allowed entanglement between microwave fields and mechanical motion to be measured using optical-quadrature measurements. While in the optical domain, photon counting measurements have been used to witness non-classical optomechanical correlations between optical pulses and phonon modes in bulk diamond via off-resonant Raman scattering, and then similarly in a nanomechanical resonator operating in the resolved-sideband regime. Furthermore, mechanical entanglement has been established using an interferometric pump–probe scheme, first between two spatially-separated diamonds with photon-counting measurements on Stokes scattered photons, and then between mechanical resonators in the resolved-sideband regime of cavity optomechanics. Moreover, continuous driving of a non-interferometric configuration with two micromechanical drum oscillators has allowed for mechanical entanglement to be established mediated by microwave fields.[1]

Neurons are mechanical harmonic oscillators

A recent essay — "Solving the 'hard problem': Consciousness is an electronic phenomenon" argues that the action potentials (APs) of neurons occur as a result of electromagnetic processes.[3] Basically magnetite crystals also known as BMNPs (Biogenic Magnetic Nanoparticles) are ubiquitous in the brain and are superparamagnetic. This means that electric flux will magnetize them as will low frequency electromagnetic radiation (EMR) that acts upon them in exactly the same way as a magnetic field.[4] Magnetizing the BMNPs in turn generates magnetic flux which opens ion channels and enables current to flow thru the membrane of the neuron. The 'spikes' generated by action potentials are therefore mediated by this complementary relationship between electric flux and magnetic flux encapsulated in Maxwell's classical equations of electromagnetism. As these 'spikes' are actually an oscillating current along the axon they will also emit radio waves at the same frequency as the current. These radio waves are universally referred to as brainwaves which likewise can magnetize the BMNPs and initiate action potentials elsewhere in the neural network. Effectively this means that all the neurons in the neural network act as harmonic oscillators.

A new research paper has detected never-before-seen waveforms in the dendrites of the pyramidal neurons in the cortex of the brain.[5] The paper argues that these waveforms are mediated by neurotransmitters in the synaptic gaps which regulates the electric flux caused by the flow of calcium and sodium ions which actually modulates the amplitude and frequency of the 'spikes'. Up to now the spikes of action potentials were thought to be typical all-or-none flow of sodium ions as the action potential propagates along the axon and were not graded in any way but this new finding indicates that in the neurons of the cortex at least, the spikes are creating waveforms that vary in amplitude and frequency in such a way as to suggest that these oscillations are transmitting information about the neurons that have fired.[6]

It is submitted that this new research confirms the theory that consciousness is generated by electronics, that the brain is an electronic device and that the brains of all living creatures are connected electronic devices. In particular it indicates that the 'firing' of specific neurons will emit brainwaves of a unique frequency that will act as a signal to other parts of the neural network. Essentially the firing of specific neurons that have been programmed for individual functions acts as a transmitter of information via brainwaves to other sections of the neural network that have likewise been programmed – 'hard-wired' - to perform supplementary processing of this signal. Essentially neurons in other sections of the brain are 'hard wired' to act as an antenna for this unique signal.

The new research above indicates that the firing of neurons is mediated by neurotransmitters at the synaptic connections in a very complex way, and it is actually the firing of the neuron itself, or a specific set of neurons, which is directly generating a signal involved in processing output. From cognitive science it is quite clear that specific sets of neurons in specific parts of the brain are engaged in performing specific functions. And the overriding question in cognitive science is how the information from these diverse firings of neurons in spatially distant parts of the brain, even from different hemispheres, can somehow simultaneously coalesce into a unified output for multiple terminals i.e. the coordination of all functions of the body and its actions in the environment at once.

It is evident that the neurons are 'hard wired' to play a coordinated role in the generation of all bodily and mental operations, and there are plenty of instances where damage to the neurons in one part of the brain will cause defects in the performance of a function, but will not necessarily extinguish it, which indicates that neurons in other undamaged sections of the brain engaged in generating that function are still operating, and it is only the final coordinated output that is in some way impaired. Also there is evidence that if a part of the brain that performs a specific function is damaged at a very early age, then neurons in another section of the brain will be 'hard wired' to perform that function. At the risk of mixing metaphors, there is evidence of considerable plasticity in the 'hard wiring' of these neurons. This indicates that the 'hard wiring' of the neurons to perform their allotted functions is in the nature of a 'programming' of the brain. But that's where the analogy to conventional computers or Turing machines ends. There are no programs written in binary code, or any other numerical symbols, that are executed by electric current running thru logic gates, which will result in the neural network generating a specific output, globally coordinated or otherwise. When the neurons themselves receive a certain signal they will fire in a very specific way, which will transmit a very specific signal via brainwaves. That is their sole function: to fire when they receive a precise signal.

Recent advances in 'mind reading' technology has established that brainwaves communicate information.[7] Brainwaves are ELF (extremely low frequency) radio waves that travel at the speed of light. New research has revealed a novel waveform in the firing of neurons in the cerebral cortex which indicates that the 'spikes' of the electric flux in the action potentials are in some way mediated or controlled so that the waveform has a precise amplitude and frequency. In other words the waveform carries information about the precise neuron or set of neurons that fired. That information is communicated instantaneously to the entire neural network via brainwaves.

It is submitted that this solves the 'binding problem' in cognitive science. How functions achieved by different parts of the brain somehow combine their output into a single conscious percept. Not only does the problem of binding occur across components of a single modality, for example the shape and color components of the visual modality, but also the problem exists for multiple modalities: how is it that a single conscious experience contains information from vision, hearing, touch, smell, and taste, all at the same time? And sensory input is all seamlessly and instantaneously merged with our body schema and motor modalities, and our thought and language modalities. The problem is that there are too many multimodal areas. There are places in the temporal lobes which receive input from all modalities (the superior temporal sulcus), and places in the prefrontal lobes which also receive input from all modalities. One candidate for the neural substrate of binding is a type of resonation or oscillation which spans all of the bound areas. The binding problem is a problem about how the different parts of a conscious state are realized in such a way that they are parts of the same state.[8] The explanation is that the neurons are entangled harmonic oscillators.

Philosophical implications of entanglement

According to Richard Feynman, one of the leading contributors to the theory "I think I can safely say that nobody understands quantum mechanics." In this article it is argued that the neurons in the brain act as harmonic oscillators that are entangled by the brainwaves that they are transmitting and receiving. Brainwaves are ELF (Extremely Low Frequency) radio waves at the very bottom of the EMR (Electromagnetic Radiation) spectrum. They travel at the speed of light and they have extremely low energy.

Although this appears to be a perfectly straight forward explanation for how the brain works we find that it is really in the nature of a 'non-explanation' because we are attempting to explain brain processes with a mathematical theory that 'nobody understands'. Obviously we can see a brain with the naked eye so the brain is a macroscopic object, but that is not a working brain. That is a brain that has been taken out of someone's skull and is no longer functional. Once we start enquiring into how individual neurons work in a living brain we find that we are immediately thrust into the deepest enigma of all about quantum mechanics theory – the question of the observer and the measuring instrument. All of brain science involves reports of an observer who has obtained results from measuring equipment. In other words that brain has what's known as a 'wave function' that collapsed at the point when the measurement was taken. The only 'information' the brain scientist has gleaned from that measurement is that the object under observation, a neuron for example, was in a certain state at the precise moment of the measurement.

Nobody knows what the 'collapse of the wave function' means. For a start it is not a real physical wave, but exists only as mathematical theory as a probability amplitude which then has to be 'squared' to obtain the actual 'probability' that a certain 'result' will be obtained. All I can do in this article is give my own interpretation of what the collapse of the wave function means and why and when it occurs, and most importantly why it requires an act of observation from a scientist. Quantum mechanics is essentially about observing the operations of microscopic objects that cannot actually be observed through the senses. So instead of actually 'seeing' the object the observing scientist sees a piece of equipment that displays a certain measurement or result. That is the irreversible point when the wave function collapses – when the observing scientist becomes 'conscious' of the result of the measurement. What was only a probability wave has become real because it has produced a 'result' on the cortex of the observing scientist. In other words the object that was measured is completely irrelevant, all that is 'real' is the result that has now irreversibly appeared on the cortex and in the consciousness of the observing scientist. There is now only knowledge of that object that was measured, that is to say 'information'. What is in the 'consciousness' of the observing scientist is critical for all quantum mechanics theory.

Once we understand that quantum theory deals with the probability of a certain result appearing in the consciousness of the observing scientist we can pass over two major enigmas relative to this discussion relatively quickly – 'wave-particle' duality and 'entanglement'. Essentially EMR (including brainwaves) is normally considered to be a wave but if you pass it thru two slits and rig up a measuring apparatus to detect which slit it went thru a result will appear in the consciousness of the observing scientist that a particle went thru one slit or the other. But if there is no measuring apparatus and no result recorded in the consciousness of the observing scientist then a wave seems to pass thru both slits simultaneously. So depending on what inquiry the experimenter is making and what methods are used information will be generated in the consciousness of the experimenter that EMR (including brainwaves) is either a particle or a wave. This says nothing about what it actually is; there is no objective reality, only what the experimenter 'knows' subjectively.

The same considerations apply for the entanglement enigma. Two entangled particles share the one wave function. When Alice takes a measurement on one of the particles and a result (say spin up) appears in her consciousness then this collapses the wave function so it may be taken as a given that the other particle will be spin down. The information has been generated in the consciousness of an observing scientist and the state of the pair of particles is determined no matter how far apart they may be. It is simply not relevant to inquire why this should be so, and it is especially not relevant to relate this back to macroscopic conceptions of time and space. The wave function collapsed at the moment the information appeared in the consciousness of Alice and that is all we are permitted to know about the matter.

We now come to this notion that neurons are harmonic oscillators. Classically we think of a harmonic oscillator as a spring with a mass at one end which stretches the spring for a certain distance. If you pull the mass down and let it go the spring will oscillate up and down about a mean position. That oscillation can be represented mathematically by a sine or cosine wave that has a frequency that can be calculated by a standard equation in classical physics. Likewise EMR can be represented by a sine or cosine wave with a certain frequency. This concept of an harmonic oscillator has been adapted in quantum mechanics to describe objects at the atomic or molecular level.

In quantum mechanics the energy for a harmonic oscillator is given by the equation $\hbar\omega/2$ where $\hbar$ (h-bar) is Planck's constant and ω is the angular frequency. This is the ground state energy for the harmonic oscillator. Without getting too technical for the harmonic oscillator there is then a positive and a negative operator that acts on the wave function to increase and decrease the energy. And here is where the word quantum in quantum mechanics becomes most significant. The energy of the harmonic oscillator is increased by a quantized amount of $\hbar\omega$. We have already seen that brainwaves are ELF radio waves that is to say their energy or frequency is at the absolute bottom of the EMR spectrum which means that the discreet energy packets or quanta by which the neuron oscillates will be infinitesimally small but they will be able to

(Published in Journal of Consciousness Exploration & Research| August 2020 | Volume 11 | Issue 5 | pp. 487-494)   57
Bartholomew, B. Y., *Quantum Theory & Consciousness*

transmit information throughout an entangled network of harmonic oscillators nonetheless. The oscillations give off brainwaves that represent discreet energy packets.

Conclusion

In previous articles I have argued that the 'binding problem' in cognitive science can only be solved on the assumption that different neurons or sets of neurons in disparate and widely separated regions in the brain are able to communicate with each other at the speed of light via brainwaves in order to generate on the cortex a unified or holistic consciousness. This paper takes that assumption one step further and argues that in fact all the neurons in the neural network are entangled harmonic oscillators emitting discreet packets of energy that enables information to merge simultaneously from diverse regions of brain to produce a holistic consciousness. Neural processes are not even limited to the speed of light. Because the harmonic oscillators are entangled the integration of information is instantaneous.

Received July 4, 2020; Accepted July 26, 2020

References

[1] Clarke, J., Sahium, P., Khosla, K.E., Pikovski, I., Kim, M.S. & Vanner, M.R. "Generating mechanical and optomechanical entanglement via pulsed interaction and measurement". *New Journal of Physics*. 22 (2020) 063001. https://iopscience.iop.org/article/10.1088/1367-2630/ab7ddd

[2] "Physicists propose how to entangle macroscopic objects using pulses of light". *PhysicsWorld*. 19 June 2020 https://physicsworld.com/a/physicists-propose-how-to-entangle-macroscopic-objects-using-pulses-of-light/

[3] Bartholomew, B.Y. "Solving the 'Hard Problem: Consciousness is an Electronic Phenomenon." *Journal of Consciousness Exploration and Research*, Vol. 11 No.1, (2020) https://jcer.com/index.php/jcj/article/view/862

[4] Stanley, S. A., Sauer, J., Kane, R. S., Dordick, J. S. & Friedman, J. M. "Remote Regulation of Glucose Homeostasis Using Genetically Encoded Nanoparticles." *Nature Medicine*. 21 (2014): 92-98. https://doi.org/10.1038/nm.3730

[5] Gidon, A., Zolnik,T.A., Fidzinski, P., Bolduan, F., Papoutsi, A., Poirazi, P., Holtkamp, M., Vida, I., Larkum, M.E. "Dendritic action potentials and computation in human layer 2/3 cortical neurons." *Neuroscience*. Vol. 367, No. 6473, pp. 83-87 (2020) https://doi.org/10.1126/science.aax6239

[6] Bartholomew, B.Y. "The Electronic Waveform of Action Potentials in the Brain." *Journal of Consciousness Exploration and Research*. (2020) 11:185-197 https://jcer.com/index.php/jcj/article/view/871

[7] Akbari, H., Khalighinejad, B., Herrero, J.L., Mehta, A.D. & Mesgarani, N. "Towards reconstructing intelligible speech from the human auditory cortex." *Scientific Reports*. 9 (2019): 874 https://doi.org/10.1038/s41598-018-37359-z

[8] Kolak, D., Hirstein, W., Mandik, P. & Waskan, J. "Cognitive Science" (2005) Taylor and Francis: New York. Kindle Edition.

Research Essay

Reconciling Consciousness with Physicalism

Ken Levi[*]

Abstract

Concerning the face-off between idealism and physicalism, both schools of thought have strengths and weaknesses. In proposing a novel approach to the question of consciousness, this article capitalizes on the strengths and rejects the weaknesses. The result is a reconciliation between the materialistic ontology of physicalism and the unitary cosmic ultimate of idealism. The "Bottom-Up/Top-Down" theory, presented here, combines two physical operations. The first is "access," how the nervous system receives and processes external stimuli, up to and including the stage of "binding." The second physical operation involves encoded signals bound together in the brain, as interpreted by a Universal Mind. How that can work as a material process is largely explained through the agency of embodied cognition.

Keywords: Ontology, metaphysics, physicalism, idealism, cosmopsychism, consciousness, embodied cognition.

1. Idealism and physicalism

In the study of consciousness, three schools of thought have emerged. These include dualism, physicalism, and idealism. Currently, dualism has largely been abandoned, leaving only two competing approaches. The idealist ontology holds that materialism is an illusion, with the only reality being consciousness, *per se* (Kozlowski, 2020, p. 387). In effect, to idealists, we live in a world of dreams. The physicalist ontology, on the other hand, maintains that consciousness is largely an illusion, a "ghost in the machine," with the only reality being material (Dennett, 1991).

Idealists, like Bernardo Kastrup (2018a, 2018b) and Itay Shani (2015), argue, in the first place, that consciousness is the only reality we know. Kastrup gives the example of an argument between Samuel Johnson and Bishop Berkeley. The Bishop, an idealist, challenges Johnson to show why he believes the physical world is real. Dr. Johnson responds by kicking a stone, and responding, "I know it thus!" He means the solidness of the rock proves it's real (Kastrup, 2018b, p. 15). Ironically, what he's actually admitting is he knows the rock exists only because he *feels* it.

The other idealist argument has to do with the "combination problem." It comes in two versions. First, how is it possible, idealists ask, for purely physical things to be conscious, since none of their components are conscious. For example, not quarks, electrons, atoms, molecules, or neurons possess anything like conscious thought. It makes no sense, therefore, to build consciousness out of mindless physical components.

---

[*] Correspondence: Ken Levi, Independent Researcher. Email: levik2016@yahoo.com

As biologist Thomas Huxley queries, "How is it possible that anything as remarkable as a state of consciousness comes about as a result of irritating nervous tissue?" (1866).

The "subject" combination problem is a dispute between two branches of idealism. Panpsychists contend that the solution to physicalism would be if every physical particle - down to the atomic and subatomic levels - possessed its own degree of consciousness. That way, these particles could combine to form higher levels of awareness. Cosmopsychists, however, respond that individual minds cannot "combine" because consciousness, by definition, is a subjective and individual property. Feelings do not aggregate.

So, building consciousness from the ground-up is problematic. Whether you build it up from inert matter, or from semi-conscious particles, either way it may not work.

Instead, cosmopsychists argue, build consciousness from the top down. If we posit a consciousness ontology, and that the "whole universe" itself is a "unitary conscious entity" (Kastrup, 2018a, p. 134), then we humans, and other sentient beings, derive our conscious abilities from above.

Physicalists, on the other hand, argue from common sense. How can 8 billion people be wrong? We are, most of us, under the impression that the material world we occupy is solid and real.

The main argument for physicalism comes from science. We know how DNA works, and how we physically evolved as a species. We know the human nervous system, and how it operates. Moreover, we observe, from implanted electrodes or FMRIs, how every part of the brain impacts a corresponding part of the body. Sentient responses, such as seeing, hearing, or feeling, can all be activated or suppressed by stimulating parts of the brain. It's only a matter of time, the physicalists say, until we learn exactly what parts of the brain produce sensation.

"It is now a canon of neuroscience that any mental experience can be associated with some specific pattern of neural firings" (Chorost, 2011, p. 33).

2. A composite theory

This essay will attempt to reconcile the best parts of both schools of thought with a composite theory that has yet to be advanced. Let's start with an analogy. We all have a multitude of household appliances. Like our brains, some of these appliances contain a bewildering nest of wiring. Yet none of them actually work until they are plugged into an outlet. Until then, they are just inert pieces of furniture. Once connected to the electric grid, however, they come alive.

That analogy illustrates what I call the "Bottom-Up/Top-Down" theory of consciousness. The bottom up part consists of what Stephen Pinker (1997) calls "access." For example, sight begins with light waves from external objects hitting our retinas, stimulating neural pathways inside out heads, and separately being processed by specialized functional areas inside our cortices. Ultimately, all sense impressions, including sight, sound, smell, etc., are tied together through a process called "binding." Every 1/40th of a second an electric wave passes over our brains and binds together all our sensory signals into a unitary frame (Blakeslee, 1998).

That part is strictly physical. One might even imagine building an android or a computer with the exact same functions. But "access" isn't enough. What we have at the binding stage is electrically encoded information. We don't have consciousness.

The top-down part consists of what I call the "Universal Mind" (UM). Once all of our stimuli are processed and encoded, the Universal Mind interprets those signals as sentient experience. The Universal Mind corresponds, in part, to what Kastrup and Shani call the irreducible cosmic ultimate. That's what we plug into.

How does it work?

1) It works physically. Contrary to the idealists, the Universe, I believe, is entirely material. Our only ontology is physical. There is no need to ditch our common sense notion that everything around us consists of solid stuff.

2) It works because we are part of the Universe. It's not just that the Universe is inside of us. We are inside of it.

3) It works through "embodied cognition." The Universe physically embodies not just us, but everything within it. For that reason, it contains not just all things, but also knowledge of all things. That knowledge is not just abstract, but also visceral. That knowledge is *existential*.

4) It works because the Universe contains embodied knowledge that the rest of do not have, and cannot have. None of us embodies the world around us. But the Universe does. For example, consider the apple. The Universe contains knowledge of its components and how they fit together. But it also contains knowledge of its existential properties, including how it looks, tastes, and smells.

5) It works because, as philosopher John Searle has observed, the properties of objects are inherent in those objects. What is it, he asks, about the raw perceptual experience that causes us to see red? His answer is, "It's part of what it means for something to be red that it's capable of causing experiences like this" (2013).

The look, smell and taste of an apple aren't figments of our imagination or constructions of our brain. They are intrinsic aspects of the thing we call an apple. Only two entities contains those aspects: the apple itself, and the Universe which embodies it.

So, like the appliance plugged into an outlet, the encoded signals in our brains are activated by the Universal Mind. That's how the existential impact of the apple is conveyed to us.

3. Objections

a) If we receive the truth of things from the UM, then why don't we all see things the same way?

Answer: the top-down part may be the same for everyone, but the bottom-up part differs. We all have different physical capabilities when it comes to the mechanical processing of external stimuli. For example, it is said that a frog "will starve to death, surrounded by food, if it is not moving" (Lettvin, et al, 1959, p. 234). A frog's visual apparatus only allows it to see animation.

b) Am I saying that the Universe, *per se*, has a "Mind;" that, in effect, the Universe is God?

Answer: Yes.

c) Where's the proof?

Answer: Proof for the existence of a Universal Mind is the topic for a separate discussion (Levi, 2019, 2020). Of course, if there is no such thing, then the present argument fails. However, the focus for this discussion is how a "Bottom-Up/Top-Down" consciousness can work.

In terms of proof, however, consciousness is like love. It's real. You know it's real, even though to outside observers, you just may be a zombie. What, then, is the best evidence for its reality? As is the case with love, the best evidence for consciousness is personal, subjective experience. The same goes for evidence of top-down consciousness. If one or two people claim it, then the evidence is merely anecdotal. But if very many people claim it, then, it becomes science (Levi, 2020).

d) Is it plausible?

Answer: Having a top-down component to consciousness is not a new idea. It's not just the idealists who espouse it. For over a billion Hindus and Buddhists throughout the world, the highest plain of existence is thought to be the Brahman, the great spirit, which dwells in each individual in the form of the Atman, the personal soul (Srinivasan, 2018, p. 402).

And, of course, "May the Force be with you" is a popular - and widely accepted - element of popular culture.

What's new here is the concept of the Universal Mind as a physical force; as a stage of human consciousness, dovetailing with our physical nervous system, and conveying existential realism.

e) How exactly can the Universe be completely physical, like us, but yet be capable of sentience, unlike us?

Answer: Embodied cognition. If, for example, I say, "I feel your pain," I don't really mean it. What I actually mean is I imagine what your pain must be like. But the Universe does, indeed, feel your pain, because the Universe is you.

f) How exactly does bottom-up "plug into" top-down?

Answer: I'm not sure. Here's what we know. Binding is the final stage in the nervous system. We have no reason to believe that what is bound together in that stage is anything but encoded signals. So, how are those signals converted into sense experiences? If it can't be explained by the physical apparatus of our brains, then, it must be something besides our brains. And that something, I hypothesize, is the Universal Mind. For one thing, we are already a part of the Universe. It doesn't have to come to us. We're already in it. For another thing, besides the object itself, only the Universe contains both objective and visceral knowledge of everything we experience.

4. Conclusion

How is this Bottom-Up/Top-Down theory different from any of the others? Answer: it explains top-down consciousness within the context of a physicalist ontology.

Unlike idealism, it doesn't throw out the baby with the bathwater. Just because we cannot currently explain - or even logically rationalize - how subjective consciousness could arise from inert matter, that doesn't have to mean the material world is fake.

Unlike physicalism, just because we cannot objectively measure or observe conscious thought, doesn't mean we are all just glorified robots.

Stephen Hawking has argued that throughout history, people have invoked the "gods" to explain things they couldn't understand. That includes earthquakes, eclipses, personal suffering, and the like. As soon as science provides a demonstrable explanation, he observes, we dispense with the gods.

As once was true of eclipses and human heartbreak, consciousness is something we cannot presently explain. Does Hawking's critique apply to this essay, as well? We will need more research to find out.

Received July 6, 2020; Accepted July 26, 2020

References

Blakeslee, S. (1998). How the Brain Might Work: A New Theory of Consciousness, (232-237). In Wade, Nicholas, Ed., *The Science Times Book of the Brain*. New York: Lyons Press.

Chorost, M. (2011). *World Wide Mind: The Coming integration of Humanity, Machines, and the Internet*. New York: Free Press.

Dennett, D. C. (1991). *Consciousness Explained*. Boston, MA: Little, Brown and Company.

Hawking, S. (2010). *Grand Design*. New York: Bantam Books.

Huxley, T. H., (1866). quoted in Block, N. (2013). The Harder Problem of Consciousness. *The Journal of Philosophy*, XCIX(8), Jan.

Kastrup, B. (2018a). The Universe in Consciousness. *Journal of Consciousness Studies*, vol. 25, nos. 5-6, 125-155.

Kastrup, B. (2018b). The Next Paradigm. *Future Human Image*, Volume 9, 2018: DOI: 10.29202/fhi/9/4

Kozlowski, M. (2020). Images, Mathematics, and Consciousness. *Journal of Consciousness Exploration and Research*, Volume 11, Issue 4, 385-391.

Levi, K. (2019). *Knowing: Consciousness and the Universal Mind*, Smashwords, https://Smashwords.com/books/view/956829 [Accessed 4 March 2020].

Levi, K. (2020). *Mind of God*, Smashwords, https://Smashwords.com/books/view/1022233 [Accessed 5 March 2020].

Lettvin, J.Y., Maturana, H., McCulloch, W.S., & Pitts, W. (1959). What the Frog's Eye Tells the Frog Brain. *Proceedings of the Institute of Radio Engineering*, vol. 47, 1940-1951.

Pinker, S. (1997). *How the Mind Works*. New York: W.W. Norton and Co.

Searle, J. (2002). The Problem of Consciousness. In *Consciousness and Language*, Cambridge: Cambridge University Press, 7–17.

Searle, J. (2013). Perception and Intentionality. *YouTube Seminar*, https://www.youtube.com/watch?v=VddLlnOZIfY [Accessed 4 June 2018].

Shani, I. (2015). Cosmopsychism: a Holistic Approach to the Metaphysics of Experience. *Philosophical Papers*, 44 (3), 389-437.

Srinivasan, R. (2018). Cosmic Intelligence, Its Manifestation and Humanity (Part I), *Journal of Consciousness Exploration and Research*, Vol. 9, No. 5, 402-416.

Exploration

Schumann Wave, Human Brain & Consciousness

Miroslaw Kozlowski[*1] & Farin Zokaee[2]

[1]Warsaw University, Warsaw, Poland
[2]IKIU, Qazvin, Iran

Abstract

When one observes the path of electron, the interference pattern/image in the two-slit experiment disappears, but why? In this paper, we discuss the interaction of Schumann wave with human brain and the role played by our consciousness.

Keywords: Image, global consciousness, two-slit experiment.

The experiments of Global Consciousness Project ("GCP") indicated that the output in Random Number Generators ("RNG") were correlated to global events (*e.g.*, Radin, 2006). The essence of the GCP was the detection of collective human brains interactions with RNG. In addition, the Earth Schuman waves may influence human thoughts and behavior. We have shown previously that both Schumann wave and brain wave have similar frequency spectra but the amplitudes of brain waves are ten times smaller.

The issue of observation in quantum mechanics is central in that objective reality cannot be disentangled from the act of observation. In the words of John A. Wheeler 1981, we live in an observer-participatory Universe. There is an impressive body of evidence that reality is non-local and undivided, thus, non-locality is a basic fact of nature, first implied by the Einstein-Podolsky-Rosen thought experiment despite the original intent to refute it, and later explicitly formulated in Bell's Theorem. This is a reality where the mindful acts of observation play a crucial role at every level. Reality, it seems, shifts according to the observer's conscious intent.

In order to put forward the classical theory of the brain waves, we quantized the brain wave field. In the model (Marciak-Kozlowska & Kozlowski, 2012) we assume that:

(i)   The brain is the thermal source in local equilibrium with temperature T.
(ii) The spectrum of the brain waves is quantized according to formula $E = h\nu$ where E is the photon energy in eV, $\hbar$=Planck constant, ν-is the frequency in Hz.
(iii) The number of photons emitted by brain is proportional to the (amplitude)^2 as for classical waves. The energies of the photons are the maximum values of energies of waves for the emission of black body brain waves we propose the well know formula for the black body radiation (Baierlein, 1998).

Besides the gravitation force, we postulated that Moon and Earth is also bounded by "consciousness field." The wave of consciousness field has the length (distance Earth-Moon) =

---

[*] Correspondence: Miroslaw Kozlowski, Prof. Emeritus, Warsaw University, Poland. Email: m.kozlowski934@upcpoczta.pl

3.4x10^5 km, velocity =light velocity=c= 3x10^5 km/s, and τ characteristic time of the order (Earth-Moon distance)/c=1.2 sec and fundamental frequency $\omega = \tau^{-1} = 0.9$s and quantum energy for fundamental frequency E=10^{-15} eV. The consciousness field influences all the phenomena on the Earth, including our brain/mind system. The brain photons may be the effect of the interaction of the consciousness with human brain. The final results of this interaction are the: *alpha, beta, delta and theta* waves.

We argue that the interaction of Schumann wave with human brain can be described by damped hyperbolic wave equation [1]

$$\tau \frac{\partial^2 \Upsilon}{\partial t^2} + \frac{\partial \Upsilon}{\partial t} = D \frac{\partial^2 \Upsilon}{\partial x^2},$$

(1)

where τ denotes the relaxation time, D is the diffusion coefficient and $\Upsilon(x,t)$ is the initial Schumann wave. Introducing the non-dimensional spatial coordinate $z = x / \lambda$, where $\lambda = \lambda / 2\pi$ denotes the reduced mean free path, Eq. (1.49) can be written in the form:

$$\frac{1}{\upsilon'^2} \frac{\partial^2 \Upsilon}{\partial t^2} + \frac{2a}{\upsilon'^2} \frac{\partial \Upsilon}{\partial t} = \frac{\partial^2 \Upsilon}{\partial z^2},$$

(2)

where

$$\upsilon' = \frac{\upsilon}{\lambda} \qquad a = \frac{1}{2\tau}.$$

(3)

In Eq. (2) υ denotes the velocity of heat propagation], $\upsilon = (D/\tau)^{1/2}$.

In the paper by C. De Witt-Morette and See Kit Fong the path-integral solution of Eq. (2) was obtained. It was shown, that for the initial condition of the form:

$\Upsilon(z,0) = \Phi(z)$ an "arbitrary" function

$$\frac{\partial \Upsilon(z,t)}{\partial t} \Big|_{t=0} = 0$$

(4)

the general solution of the Eq. (2) has the form:

$$\Upsilon(z,t) = \frac{1}{2}\left[\Phi(z,t) + \Phi(z,-t)\right]e^{-at}$$

$$+ \frac{a}{2}e^{-at} \int_0^t d\eta \left[\Phi(z,\eta) + \Phi(z,-\eta)\right]$$

(5)

$$+ \left[I_0(a(t^2 - \eta^2)^{1/2}) + \frac{t}{(t^2 - \eta^2)^{1/2}} I_1(a(t^2 - \eta^2)^{1/2}) \right]$$

ISSN: 2153-8212   Journal of Consciousness Exploration & Research   www.JCER.com
Published by QuantumDream, Inc.

In Eq. (5), $I_0(x)$ and $I_1(x)$ denote the modified Bessel function of zero and first order respectively.

Let us consider the propagation of the initial Schumann wave with velocity v', i.e.,

$$\Phi(z - v't) = \sin(z - v't) \tag{6}$$

In that case, the integral in (5) can be computed analytically, $\Phi(z, t) + \Phi(z, -t) = 2sinzcos(v't)$ and the integrals on the right-hand side of (5) can be done explicitly ,we obtain:

$$F(z,t) = e^{-at}\left[\frac{a}{w_1}\sin(w_1 t) + \cos(w_1 t)\right]\sin z, \qquad v' \geq a \tag{7}$$

and

$$F(z,t) = e^{-at}\left[\frac{a}{w_2}\sinh(w_2 t) + \cosh(w_2 t)\right]\sin z, \qquad v' < a \tag{8}$$

where $w_1 = (v'^2 - a^2)^{1/2}$ and $w_2 = (a^2 - v'^2)^{1/2}$.

In order to clarify the physical meaning of the solutions given by formulas (7) and (8), we observe that $v' = v/\lambda$ and w_1 and w_2 can be written as:

$$v_1 = \lambda w_1 = v\left(1 - \left(\frac{1}{2\tau\omega}\right)^2\right)^{1/2}, \qquad 2\tau\omega > 1,$$

$$v_2 = \lambda w_2 = v\left(\left(\frac{1}{2\tau\omega}\right)^2 - 1\right)^{1/2}, \qquad 2\tau\omega < 1, \tag{9}$$

where ω denotes the pulsation of the initial Schumann . From formula (9), it can be concluded that we can define the new effective brain wave velocities v_1 and v_2. Considering formulas (8) and (9), we observe that brain wave with velocity v_2 is very quickly attenuated in time. It occurs that when $\omega^{-1} > 2\tau$, the scatterings of the heat carriers diminish the thermal wave.

It is interesting to observe that in the limit of a very short relaxation time, i.e., when $\tau \to 0$, $v_2 \to \infty$, because for $\tau \to 0$ Eq. (2) is the Fourier parabolic equation. It can be concluded, that for $\omega^{-1} > 2\tau$, the Fourier equation is relevant equation for the description of the brain waves . For $\omega^{-1} > 2\tau$, the scatterings are slower than in the preceding case and attenuation of the thermal wave is weaker. In that case, $\tau \neq 0$ and v_1 is always finite:

$$v_1 = v\left(1 - \left(\frac{1}{2\tau\omega}\right)^2\right)^{1/2} < v \tag{10}$$

For $\tau \to 0$, i.e., for very rare scatterings $v_1 \to v$ and Eq. (2) is a nearly free t l wave equation. For τ finite the $v_1 < v$ and consciousness wave propagates in the medium with smaller velocity than the velocity of the initial thermal wave.

Considering the formula (2), one can define the change of the phase of the initial consciousness wave β, i.e.:

$$\tan[\beta] = \frac{a}{w_1} = \frac{1}{2\tau\omega} \frac{1}{\sqrt{1 - \frac{1}{4\tau^2\omega^2}}}, \qquad 2\tau\omega > 1. \tag{11}$$

We conclude that the scatterings produce the change of the phase of the initial Schumann wave. For $\tau \to \infty$ (very rare scatterings), $\tan[\beta] = 0$.

Let us consider the two-slit experiment. When large molecules are heated by laser light during the two-slit experiment the interference image disappeared. In our interpretation, it means that $\tan[\beta] \neq 0$

Received May 30, 2020; Accepted June 13, 2020

References

Schild R. Cosmology of Consciousness, Quantum Physics & Neuroscience of Mind, Cosmology Science Publishers, Cambridge, 2012.

Baierlein R. Thermal Physics, Cambridge University Press, 1999.

Marciak-Kozlowska J, Kozlowski M. Heisenberg's Uncertainty Principle and Human Brain. NeuroQuantology 2013; 11(1): 47-51.

Kozlowski M, Marciak-Kozlowska J. Brain Photons as the Quanta of the Quantum String. NeuroQuantology 2012; 10(3): 453-461.

Durrer R. The Cosmic Background. Cambridge University Press, 2008.

Levy-Leblond J, Balibar F. Quantics. Elsevier Publisher, 1990.

Radin, D. I. (2006). Entangled minds: Extrasensory experiences in a quantum *reality*. New York: Simon & Schuster (Paraview Pocket Books).

Kandel E R et al. ,Principles of Neural Science , Fifth Edition, McGraw-Hill, USA, 2012

Marciak-Kozłowska, J, Kozłowski M, Extrasensory perception phenomena, Lambert Academic Publishing, (and references therein), 2016.

Marciak-Kozlowska, Kozlowski M, Binding energy of human brain, Journal of Cosciousness Exploration and Research, 2017.

Exploration

Hypno-Channelings:
A New Tool for the Investigation of Channeling Experiences

Luciano Pederzoli[1], Elena Prati[1], Nadia Resti[1], Daniela Del Carlo[2]
& Patrizio Tressoldi[*3]

[1]EvanLab, Firenze, Italia
[2]S.A.M.O. - Scuola di Counseling Olistico, Avenza (MS), Italia
3Science of Consciousness Research Group - Dipartimento di Psicologia
Generale, Università di Padova, Italia

Abstract

This study describes for the first time a procedure to channel and interview various purported discarnate Entities obtained by initially inducing the participants into an Out of Body Experience via hypnotic suggestions. We report the conversations with seven different Entities, three of which were done individually and four in group sessions, channeled by different participants with no previous channeling experience. This procedure, we named Hypno-Channeling, allows to contact and interview with easy various Entities and it represents a good tool in order to determine whether the acquired information derives from a discarnate Entity, the channeler or the interviewer.

Keywords: Channeling, hypnosis, out-of-body experience, phenomenology.

Introduction

Gospel of Luke 24:27
Then beginning with Moses and with all the prophets,
He explained to them the things concerning Himself in all the Scriptures.

Quran, chapter 96 (Al-Alaq), verse 1-5
Proclaim! (or read!) in the name of thy Lord and Cherisher, Who created-
Created man, out of a (mere) clot of congealed blood:
Proclaim! And thy Lord is Most Bountiful,-
He Who taught (the use of) the pen,-
Taught man that which he knew not.

Channeling is a practice in which an individual (a channeler), usually willingly, enters into a non- ordinary mental state and communicates with one or more intelligent 'Entity' that speaks

---

[*]Correspondence author: Patrizio Tressoldi, Science of Consciousness Research Group, Dipartimento di Psicologia Generale, Università di Padova, Italia. E-mail: patrizio.tressoldi@unipd.it

ISSN: 2153-8212

and acts through his/her mind and body. During such a state, the person acting as the channeler may or may not be consciously aware of what the Entity is saying or doing.

Whereas channeling experiences have been known since the origin of all the major religious traditions, very few scientific investigations have been carried out to date (see Stolovy, Lev-Wiesel, & Eisikovits, 2015; Wahbeh, Carpenter, & Radin, 2018; Wahbeh & Radin, 2018; Wahbeh, Cannard, Okonsky, & Delorme, 2019; Jooyce, Wahbeh, Delorme, Okonsky, 2020). Among the main issues related to this phenomenon is the critical question about the origin of the channelled information and specifically: *"Is this information transmitted by a discarnate Entity of a different physical/spiritual level or does it simply come from the mind of the channeler?"*

In modern times this practice is usually restricted to mediums and clairvoyants, which often 'channel' – for payment or sometimes for free – a client's deceased person. Other times the channeler, either through a specific practice or sometimes spontaneously, senses an Entity requesting the medium's mind or body to communicate something to him/her or others.

The following are extracts of descriptions of their experiences from two spontaneous channellers.

Sanaya & Duane (1987):

> My experience of trance varies, depending on the information I am bringing through. I go into a very deep trance when channeling information for books and relaying esoteric information of universal knowledge. When I am channeling for other people, my trance is lighter, for it does not take the same amount of Orin's energy to transmit this kind of information.

Patrizia Vale (2018):

> On advice from D. Baker I began keeping a 'spiritual diary' after my meditations, in which I wrote general questions such as: "What does Man add to God's perfection? Why did the ONE need duality?", and other personal ones. To my great surprise I began receiving answers, at first succinct and then very articulate, all in English.

> This was my first contact with a Guide – I was told – I had met during an incarnation in the USA. These handwritten contacts continued almost daily until 2000. From this guide I learned many things about my past, got much invaluable advice about my personality and was forewarned that I would be contacted by a 'higher Entity'.

> Indeed, as I wrote in the introduction to 'Let the soul speak', in the summer of 2000 – I had just bought a computer I barely knew how to use – I felt the need to write the first page, which began like this: "When we work for the soul we are always assisted by an Angel and many Guides...". Was it a voice? A force? To my amazement the dictation

continued and after a month my first book about the soul was created, containing 12 chapters.

Even now I still can't describe what I feel as I write; I know I need to keep my mind empty and simply be a receiving antenna. I must not interfere with my mind, but have total faith in it. I am not in a 'trance' state, I'm fully conscious and can decide when I'm ready and when to stop writing.

From a study by Stolovy et al (2015) and one by Wahbeh et al (2018) respectively, we can read about further experiences obtained from individual interviews of 20 expert Israeli channelers and five American ones working as a group.

These two studies show that the channelings were positive and controlled, attesting that these dissociative experiences are not expressions of mental psychopathology, as also confirmed in the study of Wahbeh & Radin (2018).

Since channeling is not usually replicable at any moment, the channeler, if capable, could telepathically access the questioner's mind. We therefore come to the fundamental question: *"Could the information supplied by the non-physical channeled Entity perhaps be a product of the channeler's or the questioner's mind?"*

A reliable answer to this question requires repeatedly and easily channeling the particular Entity and subjecting it to a long series of well-prepared questions, the answers of which will only be of specific interest if they are not part of the knowledge/beliefs of the channeler or the questioner.

In the procedure paragraph that follows, we will describe an innovative method which allows selected participants to be first induced in a controlled Out of Body Experience (OBE) by specific hypnotic suggestions and afterwards to be brought in contact with disincarnate Entities interested to be channeled. In this controlled channeling condition, the hypnotist can interview the channeled Entity asking all questions he is curious about and try to understand whether the information he is receiving are really conveyed by a "third" part and not from the channeler itself.

Methods

Participants

The participants were nine adults – eight females and one male – ranging in age from 52 to 68 years with no self-referred physical or mental pathology. Those who did the individual sessions, all females, we will call E, DD, and N, while those in the group session will be called A, An, M, D, P and DC, as well as DD who also did individual sessions.

On the Dissociative Experience Scale – II, seven participants (M, A, An, D, P, DC and DD) reported a score greater than 30, which is the threshold of possible psychopathology. However, their professional and personal lives are totally free from clinical mental disorders as declared by them and witnessed by the hypnotist who knew each of them from long time.

All participants signed an informed consent form containing a description of the entire procedure, its aims, and how to voluntarily interrupt the experience at any time.

Each participant had accrued experience in hypnosis and hypnotically induced OBEs varying from 2 to 10 sessions with the same hypnotist, but none had ever experienced a channeling.
The hypnotist was an adult male with more than 20 years of experience in research hypnosis.

Procedure

Contact with the channeled Entity occurred according to the following four stages: Relaxation, hypnosis induction, OBE induction, contact with the channelled Entity.

The characteristics of these stages and the procedures used are described in detail in Pederzoli and Tressoldi, (2018) . In short, the purpose of the relaxation stage, with a maximum time of four-five minutes, is to remove the subject's attention from normal bodily sensations, which in this situation are considered unwanted background noise. The pre-hypnotic phase, with a maximum duration of four-five minutes, is based on the induced visualization of a relaxing environment totally free of anxiety but rich in rapidly alternating sensory stimuli. A final stage, with a maximum duration of three minutes, is added in which sensory stimulation continues further, then suddenly stops when the subject is led to an armchair which envelops him/her like a cloud, causing all physical sensation to cease, thus marking the true OBE stage. The attainment of this state of consciousness was directly asked to the participants who confirmed they felt themselves detached from their physical body.

It must be specified that the time required to reach the OBE induction stage has shortened with repeated sessions, down to a few tens of seconds.
Contact with the Entity occurred by simply asking the OBE subject to make him/herself available for contact and then wait attentively.

At this point the hypnotist would ask if the Entity was ready for a dialogue and if the answer was positive (we never encountered a refusal), the conversation would proceed. Apart from the first contact when, obviously, it was impossible to predict which Entity would appear, a list of questions was prepared by the hypnotist before each subsequent encounter.
It is important to take in account that the channelers didn't know the questions before their channelings making impossible to retrieve the responses in advance.

Audio recordings were made of all interviews so they could be listened to again at leisure to organize the answers and prepare further questions. Some examples of complete interviews are available at https://figshare.com/articles/Channelings/6984251.

In addition to single participants, the OBE induction technique for channeling was also applied to groups, once with four and another with six female adults and one male adult, all without prior channeling experience. All participants already had some experience in OBE induction by hypnotic suggestion with the same hypnotist, and in a group auto-induction. Essentially the procedure for OBE induction in a group is the same as for an individual, except that there are more participants and they are all present in one room, either sitting or lying on the floor in close proximity.

When all participants declared having reached the OBE state, the hypnotist would simply instruct: "*Now that you are all in an OBE state, hold hands and convey the intention to contact non-physical Entities.*" The complete transcription of two group channellings are available at: https://figshare.com/articles/Channelings/6984251

At the end of each channeling session, all audio recordings were transcribed and were made available to the channelers in order to show them the information they channeled. Furthermore, they were requested to indicate what information they considered new and unexpected for them.

Results

Individual Channellings

The entity channelled by E described itself as 'Agarthian' and the channeling (to date more than nine hours in total) was indirect, in that E acted as the 'interpreter', translating, so to speak, the questions and the Entity replies. As efficient as it is, this type of channeling can lead to a loss of information in the double 'translation' process; nonetheless the excellent replicability of these same channelings and the quality of answers allowed a detailed study of this Entity and its knowledge. The information obtained from this channeling is available in the file titled 'Agarthian Information ' at https://figshare.com/articles/Channelings/6984251. The pink text in this file indicates information unknown to both the questioner and the channeler.

Since, however, that Entity seemed somewhat reluctant to disclose certain information and in order to cross check his information, from the 24th April 2017 it was possible to channel still indirectly another "Agarthian' Entity with DD (to date 2 hours and 23 minutes in total), calling itself Antares and who, as well as disclosing some unexpected information, stated that it knew the other 'Agarthian' and his disinclination to impart information of interest to his despised 'humans'. The information obtained from this channelling is available in the file titled 'Antares' Interview" at https://figshare.com/articles/Channelings/6984251.

From these channellings we obtained information on the following subjects:

- Differences and relationship between the Physical, Subtle, and Psychic Bodies;
- The afterlife and different lives;
- The structure of reality;
- Mediumship;
- Mind-matter interaction;
- Medicine;
- Information about Agarthians and other populations.

The third single participant, N, proved to be an excellent channeler. When asked to make herself available for contact with non-physical Entities who appear trustworthy and with a reassuring aura, N reported seeing a very bright white light in front of her. Within it was a luminous elderly man with long white hair and a long white beard. When queried, "He" (as we referred to it) replied that he was available and had appeared exactly for this reason. He also said he knew what we were doing and looked upon us benevolently.

From the second contact onwards, however, this Entity appeared to N as only energy, in that he was – according to "Him – composed of 'thought', meaning will/intent – in this case intent for contact – and 'energy', meaning 'thought density', and not as a product of force multiplied by distance (as in physics).

From the 21st October 2017 until this writing, more than 29 hours of actual channeling were collected with this particularly interesting Entity. Furthermore, after a few hours of indirect channelling, it was possible to switch to regular direct channeling, which is to say, speaking directly with the Entity, with the channeler having more frequent and deeper spontaneous amnesia regarding what occurred during the channeling.

Here is a description of the experience from the channeler herself:

There were three stages: the first one, very short, was the 'driving school' as the hypnotist called it. I had no difficulty following the instructions, even though they were a bit odd; there was a strong component of 'judgement' which sometimes came through in the answers, but I had neither unpleasant sensations nor strong resistance that could have interfered with the process.

The second was the stage in which the instructions had been learned: I was very relaxed in the experience. The 'rational' part was becoming less present, as my limbs were becoming more lethargic, especially my arms. In short, my body was resistant.

The third stage began for me when my conscious part became like a spectator and was no longer involved. I could summarize it by saying it was like watching a performance in which part of me was taking part in it and another part was a spectator. This more or less coincided with the moment in which LP started talking directly to the one we called "He".

It was "He", with respect to questions the hypnotist asked about my physical difficulties, who gave us very practical suggestions on managing them. Suggestions I followed and in fact the problems gradually disappeared, such that channeling sessions lasting longer than an hour posed no problems. Now my physical reactivation is in the order of a few minutes.

This description refers to the changes I noticed during the channeling, but I also noticed differences compared to normal even in the extra-channeling periods. At first, in what I called the 1st stage, after the sessions I remembered the subjects we discussed quite well, whereas by the second stage I only remembered some things, and I only remembered the later ones after listening to the recording.

In the third stage, where we are now, I don't remember what was discussed during the channelings and I retrieve it when I listen to the recording.
I often deliberately wait five or six days after the channeling sessions before listening to the recording and in the listening I am amazed by the answers to questions, because I really couldn't answer them with what I know, and I listen with great curiosity about what "He" says. Even now I'm still not used to the consistency with which many concepts are stated, repeated and expanded on.

By now for me the channeling sessions are normal; I have been living this experience continuously for about a year. After overcoming stages 1 and 2 I feel very alert and mentally resistant, and I noticed an improved intuitive capacity. I can't be certain that this is due to the sessions, so I share it as a personal feeling, but I emphasize that it's a rather distinct sensation.

This procedure has allowed rather detailed investigation into the channelled personality and his disclosures, that range from characteristics of our identity to life after death of the Physical Body, from the structure of three-dimensional reality to a multi-dimensional one, from how to use our mind's potential to the story of humanity. The complete content of the interviews is detailed in the file titled 'Contents of dialogues with "Him"' available from https://figshare.com/articles/Channelings/6984251

The following is a short excerpt from the interview conducted on the 22nd April 2018:

H (hypnotist) – Physical Bodies can have intense interactions among themselves: can Subtle Bodies do the same, or not?

"He" – Yes, the Subtle Body is very connected to the Physical: if the Physical Body is in a punch-up, the Subtle Body is not giving caresses, but the Subtle Body is not just a casing for the Physical Body, it's another level, so the opposite is also true, in that if the Subtle Body enters an elevated state, the Physical Body does too, like in an OBE.

H – If we arrange to meet in an OBE, such as while we sleep, does the meeting occur between the Subtle or the Psychic Bodies?

"He" – The Psychic Bodies.

H – How is it, then, that the meeting is in an environment that appears real, such as in a room that doesn't actually exist, but there's a sofa on which the two sit exactly as they would with their Physical Bodies? Is it just a representation, or something more?

"He" – It's a representation, a crutch that you all need.

H – Obviously Psychic Bodies can interact among themselves – what can they do?

"He" – Lots of things (chuckles), depending on the reason: they can be very 'strong', in the usual meaning of the term, but their interaction is rather complex.

H – Is it at the level of information exchange or other levels?

"He" – Information exchange could represent a lower level rather than a simpler one. Psychic Bodies are at a very high level, almost at the energy exchange level.

H – So it's a transmission of deeper concepts, and not just information...

"He" – Yes, even at that level individuality is very blurred.
And here is an extract from the interview conducted on 19th August 2018:

H. – To what degree are you independent of my and N's cultural conditioning and beliefs?

"He" – I've been hoping you would ask me that. I can't say I'm totally independent, also because if I were totally independent, I wouldn't be able to make contact. To do so I need to connect to your intuitions. Intuition is the most correct concept and without it there would be no contact. Obviously, and you would have noticed, our interaction is gradually changing, but it must happen slowly and is dependent on both of you. What I say doesn't depend on you; how I express it depends on your conventions, culture, and ability.

H. – In fact, your statements sometimes coincide with what I think and other times they don't, and for me those that don't agree with either my or N's convictions are the most credible because when something you say coincides with my point of view, I wonder if it was me transmitting that idea.

"He" – I know very well that you haven't always believed everything I've told you.

Group Channelings

The subjects covered by the two Entities are concentrated on the differences between our universe and others. The full translation of these two channellings is available in the file titled 'Group channelings while in hypnobe' from https://figshare.com/articles/Channelings/6984251

Discussion

The achievement of an OBE state via hypnotic induction has allowed some very interesting channelings with participants who had never previously undergone such an experience.
After more than 40 total hours of channeling with nine different channelers and seven different channeled Entities, the repeatability of this procedure is therefore not just scientifically approachable, but can also be used for new and worthwhile research into a heretofore unexplored field.

We advise those who seek to use our procedure to be very mindful of the goal of these experiences and the personal characteristics of the channelers in order to avoid dissociative effects or possessions which can cause harm to their mental health.

Among the many advantages offered by this procedure with respect to voluntary and involuntary channeling, we emphasize the fact that it allows the 'translating' of information given by various Entities into words and concepts familiar to the questioner, even though many times these same Entities had difficulty finding words to express what they wanted to say, for example:

H – If I understand correctly then, 'my' Psychic Body has multiple identities – of which LP is a part – that make up a 'lives plan', but in turn is part of a 'higher' plan of which various Psychic Bodies are a part: we could perhaps call it a 'Super Psychic Body'.

"He" – It's better to call them 'branchings'. What I'm doing – my activity – is at the level of your current Psychic Body: it's part of a bigger plan and acts on your Psychic Bodies. The situation is not as hierarchical as you make it sound, it's more 'fluid'. Your situation, with different interacting Psychic Bodies, is not very common: I can, in a way, 'see' your branchings and when I mentioned many lives (the exact number is not important) I was referring to the overall plan, which in your case is rather complex.

Other times though they clearly stated that certain things for now cannot be understood, for example:

> H – Sometimes, when speaking of human evolution so far, you used words like 'maybe' or 'probably': is it limitations of your knowledge or in the explaining?

> "He" When I used the word 'maybe', it was appropriate at the time because I was taking into account the range of possibilities for humanity's development and I chose what was needed for the answer, but there could have been others.

> H – Are there other recordings like those of our past, potential, or parallel history?

> "He" – Depends on what we talk about: we spoke about many evens on this Earth, but it's a partial conversation. We could risk being 'diverted', because even for questions about Earth's history we need to adhere to a type of boundary, or we could truly lose ourselves. It's not that I, as an Entity, don't know: from where I am, I can easily access that information, but it wouldn't be useful.

The above excerpts underline the importance of practice for the channelers to reduce their cognitive filters and hence allow a direct conversation between the interviewer and the Entity.

Do we have enough evidence to prove that the information obtained is from Entities and not from the channeler and interviewer/hypnotist minds?

Those who conducted and participated in these experiences firmly believe so, but it is certainly not enough to convince those who think that all the information received would have been drawn from the memory – even if unconsciously – of the channelers and/or hypnotist. The bottom line is that until now, the information received does not contain predictions of events or knowledge that can be verified in the immediate future.

We can definitely exclude the possibility that the information was obtained by the telepathic connection between the channeler and the questioner's conscious mind, in that it would make no sense for the questioner to still be collecting information from the channelings if it were only what was already known.

Conclusions

Whether or not the information received was simply obtained from the channeler remains an open question. Our channelers ruled this out categorically after having read the contents, but although they had the interviewer's total trust, we cannot entirely exclude this possibility even though it seems highly unlikely considering the variety of subjects covered.

We believe that, for the time being, the best proof of the source of the information lies in independent repetition by different hypnotists and channelers.

If these replications will confirm that the information received comes from non-physical Entities, we believe our procedure may represent a true and compelling possibility of communicating with Entities living in a different dimension to ours. If not, it would still be very interesting to investigate how the channelers can obtain such information that would otherwise be unavailable in an ordinary state of consciousness.

Acknowledgments: Part of this research was funded by the Helene Reeder Memorial Fund. We also acknowledge the contribution of Dr. Cinzia Evangelista for English revision.

Received June 22, 2020; Accepted July 25, 2020

References

Joyce, A., Wahbeh, H., Delorme, A. & Okonsky,J. (2020). A qualitative exploratory analysis of channeled content." *Explore,16,4,231-236;* doi.org/10.1016/j.explore.2020.02.008

Pederzoli, L., & Tressoldi, P. E. (2018). A Guide for OBE Induction. *SSRN Electronic Journal.* https://doi.org/10.2139/ssrn.3148432

Sanaya, R., & Duane, P. (1987). *Opening to Channel: How to Connect with Your Guide.*

Stolovy, T., Lev-Wiesel, R., & Eisikovits, Z. (2015). Dissociation and the Experience of Channeling: Narratives of Israeli Women Who Practice Channeling. *International Journal of Clinical and Experimental Hypnosis*, *63*(3), 346–364. https://doi.org/10.1080/00207144.2015.1031555

Vale, P. (2018). Pubblicazioni - Centro Yoga Shakti. Retrieved August 19, 2018, from http://www.centroyogashakti.com/pubblicazioni/

Wahbeh, H., Cannard, C., Okonsky, J., & Delorme, A. (2019). A physiological examination of perceived incorporation during trance. *F1000Research*, *8*, 67. https://doi.org/10.12688/f1000research.17157.1

Wahbeh, H., Carpenter, L., & Radin, D. (2018). A mixed methods phenomenological and exploratory study of channeling. *Journal of the Society for Psychical Research*, *82*(3), 129–147.

Wahbeh, H., & Radin, D. (2018). People reporting experiences of mediumship have higher dissociation symptom scores than non-mediums, but below thresholds for pathological dissociation. *F1000Research*, *6*. https://doi.org/10.12688/f1000research.12019.3

Exploration

Mind & the Expressions of Mind

Holly M. Pollard-Wright[*]

Abstract

This article presents a theory of mind and the expressions of mind by way of an explanation of how our universe and 'consciousness 'came about but it does so in a non-dogmatic way. It represents science as a melding and correlation of philosophies from a multitude of disciplines with a logical framework whereby terminology may be shared with other theories, although meanings may differ. Accordingly, attributes from theories have been used which have symbolic meaning and include energy, matter, space and structure. These represent expressions of mind according to this theory and thus are phenomenological representations. Terms are defined throughout the text and illustrations are provided which may serve to further clarify concepts.

Keywords: Mind, consciousness, energy, matter, dark, normal.

Introduction

Pure awareness, pure mental, and mental images represent three fundamental expressions of mind and thus as states of mind. Accordingly, the relationship between them may be understood in the relationships which bring them together and will be discussed going forward by way of the phenomena that represent these fundamental expressions of mind: Dark energy representing the pure awareness state will be denoted dark energy/pas; focal points of dark matter (FPDMs) representing the pure mental state will be denoted FPDMs/pms; and normal matter representing the mental images state will be denoted normal matter/mis. Mind without beginning or end can express itself. In doing so three fundamental realities are manifested which will be discussed in the text: Inconceivable reality, unobservable reality and observable reality (see Figure 1):

---

[*] Correspondence author: Holly M. Pollard-Wright, Independent Researcher. E-mail: holly.pollard-wright@wildridevet.org

Mind
Isotropic Space

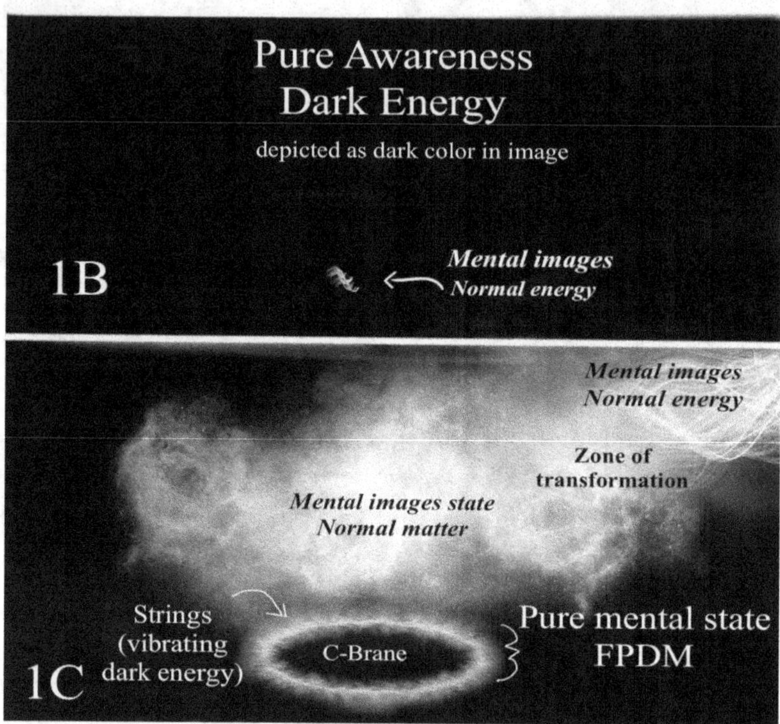

Figure 1

1A. Mind may be simply 'inconceivable' represented by isotropic space and can express itself.

1B. Mind may express itself as pure awareness represented by dark energy, but where normal energy representing mental images is present but not expressed. Accordingly as two *non-interacting* energies (i.e., dark energy and normal energy) in a structured relationship, the nature of mind manifests as 'unobservable reality' and a macrocosm.

1C. Mind may express itself as pure awareness and mental images with the potential for transformation represented by dark energy and normal energy respectively.

Therefore as two *interacting* energies (i.e., dark energy and normal energy) in a structured relationship, mind, in addition to pure awareness and mental images, expresses itself as pure mental represented by focal points of dark matter (FPDMs). Accordingly as three entities (i.e., dark energy, FPDMs and normal matter) the nature of mind manifests as 'observable reality,' in a macrocosm, a microcosm and intelligence. In the macrocosm where the cause is substantial, unconscious intelligence emerges whereby the cause and effect communicates the nature of mind in a similar way. This is represented by FPDMs/pms with the intelligence of primordial consciousness emerging from the universe's dark energy/pas. In the microcosm where the cause is a cooperative condition, conscious intelligence emerges whereby the cause and effect communicates the nature of mind in ways which may be quite dissimilar. This is represented by embodied FPDMs/pms with primordial consciousness interacting with normal matter/mis with the intelligence of conscious events and by way of their *relationship* conscious intelligence emerges (i.e., 'consciousness').

Accordingly the 'dimensions of unconscious intelligence' (i.e., 'dimensions of primordial consciousness') of FPDMs/pms and normal matter/mis with the intelligence of conscious events are connected. They are connected by relationship whereby particular events act as stimuli termed Feelings of Knowing (FOKs) (discussed in the text), the nature of mind is spontaneously communicated as conscious intelligence (i.e., 'consciousness')(1). Therefore conscious intelligence 'consciousness' includes subjective experience, but always in relation to the close analogues 'primordial conscious dimensions' and variation in the intelligence of conscious events. In this way, reality and the expressions of mind are synonymous.

This theory may thereby represent a useful tool for explaining how mind with its three states functions, arranged in what may be a coherent logical scheme of cause and effect which may occur during historic events:

- **'Inconceivable' reality prior to the formation of the macrocosm**
  Mind, inconceivable and represented by isotropic space with properties that were uniform thus had features which would be impossible to measure. Accordingly isotopic space represented a certain continuity of mind and due to its immeasurable potential for transformation, expressed itself (see image 1A).

- **'Unobservable' reality and formation of the macrocosm prior to the Big Bang(s)**
  Mind was not 'created' as if it came into existence, exists, and will cease to exist. Mind with infinite *potential* and in an essential way, expressed itself as a distinct state, pure awareness and an indistinct state, mental images. Accordingly isotropic space as a single fundamental entity representing mind manifested a macrocosm with dark energy as a homogeneous expanse representing the pas and normal energy representing mis. Dark energy/pas was a much larger mass density as compared to normal energy/mis (2). Therefore what came into existence was *transformation* whereby the former is an extension of the latter (see image 1B)

Pure awareness as a state of mind lacks a discernible pattern and thus cannot be mentally grasped as data. Accordingly the pas may be characterized as having an 'illuminate' nature beyond all concepts of existence and non-existence, subject/object reference, and a sense of an independently existing 'local' self. Dark energy representing the pas is an unknown influence causing the rate of expansion of the universe to increase and time in the macrocosm to be a homogeneous measurement that fills the whole of space-time.

- **'Observable' reality, the Big Bang(s), the formation of the microcosm and intelligence**
  The nature of mind manifests as 'observable reality' in a macrocosm, microcosm and intelligence due to mind having expressed itself as three fundamental states (i.e., pas, pms, mis). Accordingly these states can communicate the nature of mind uniquely but only due to their relationship with each other, or mind itself and where particular movement exists between the three states as action, reaction, cause and effect. Mind with its three states and the manner by which they currently function will be the focus of the remainder of this article and by way of the historic events: 'Observable' reality, the Big Bang(s), the formation of the microcosm and intelligence.

Relationships that began with the Big Bang(s)

The Big Bang(s) may represent the start of this cycle which is one of an infinite number of cycles of interactions. (3). Accordingly in the macrocosm, dark energy/pas is transformed into compacted multi-dimensional space by way of filaments of vibrating dark energy (i.e., open strings and closed strings)(4). This was due to it being a much larger mass density than normal energy/mis and the possibility that normal energy/mis would have been 'swallowed up 'rather than engaging in relationship with dark energy/pas. Therefore dark energy/pas morphed and manifested perfect geometric shapes of vibrating strings of dark energy circles emerging as FPDMs/pms. A circle of dark energy/pas demarcated by vibrating dark energy strings (i.e., open, closed) is a membrane of compacted multi-dimensional space termed a C-brane (5) (see image 1C). Accordingly a C-brane is an 'object' which generalizes the conception of a round two-dimensional shape to the higher dimensions of dark energy/pas. It serves as a 'storehouse 'for strings (i.e., open, closed) which are not destroyed, but rather are converted from open to closed or closed to open (6).

Two types of strings (i.e., open, closed) as vibrating dark energy/pas when in a circle create a C-brane. This creates limits, consequence and duality as relates to the transformative properties of the universe's dark energy/pas and compacted multi-dimensional space (i.e., C-brane). Accordingly FPDMs/pms represent a mechanism by which the non-local gravitational field of dark energy/pas in the macrocosm becomes local (7). The emergence of FPDMs/pms due to particular action by dark energy/pas as the morphing of vibrating strings (i.e., open and closed) brings about a change in universal local properties. This as a consequence causes the transfer of dark energy/pas (8). Therefore a FPDM/pms is increased by way of an amount of concentrated dark energy/pas whereby some of its *closed strings* change vibrations from their ground state to their first excited state (open and closed strings share the same ground state when vibrating with a minimal amount of energy and the lowest mode of vibration) (9). Consequently a FPDM/pms generates momentum as gravity/awareness in motion and this pulls normal energy/mis towards it (see image 1C).

The outcome of this event is transformation. Accordingly normal energy/mis not previously expressed is expressed as normal matter/mis with the intelligence of conscious events that embodies each FPDM/pms. Therefore gravity in the macrocosm is not a force but rather a consequence of the curvature of homogenous dark energy/pas that transforms into FPDMs/pms. These events 'highlight' the transformative properties of the states of mind (i.e., pure awareness state [pas], pure mental state [pms] and mental images state [mis]) which may occur, but not without consequence due to their interdependent relationship. Accordingly FPDMs/pms are the means by which dark energy/pas as a homogenous expanse with attributes that are non-local in the macrocosm are transformed into local qualities in the microcosm. This allows it to engage in relationship with normal matter/mis in innumerable ways.

When embodied by normal matter/mis, FPDMs/pms are no longer in the macrocosm but instead are located in the microcosm. FPDMs/pms are the dark energy/pas of the macrocosm folded into a discrete volume of normal matter/mis. Accordingly they may have a size comparable to a hypothetical length-scale with no smaller length possible which has a definite meaning, or may be modeled as having no width. Each FPDM/pms has its own position on the universe's dark energy/pas substrate layer in relation to all other FPDMs/pms. Therefore the microcosm represents a set of circumstances characterized by innumerable relationships of FPDMs/pms embodied by normal matter/mis. Accordingly the microcosm 'encapsulates' the features of something larger (i.e., dynamical macrocosm/relationships between the states of mind) (10). Normal energy/mis is most commonly in its form as normal matter/mis which embodies FPDMs/pms. The momentary occurrence where normal energy/mis is in the form of normal energy/mis is brief and undetectable.

Configuring Relationships in the microcosm

In the microcosm gravity is too weak to configure normal matter/mis that embodies FPDMs/pms. Accordingly in the microcosm the fundamental force associated with configuring relationships is electric and magnetic fields (i.e., electromagnetic force) generated by the vibrations of strings (i.e., open, closed). This organizes normal matter/mis into an integrated unit.

Embodied in the microcosm while interacting with normal matter/mis, open and closed strings of FPDMs/pms vibrate in their excited state and create forces (i.e., contact forces) as electromagnetism. This causes normal matter/mis to bend and give way and with stresses upon it turns 'outside-in' or 'inside-out' (11). Over billions of years of relationships where FPDMs/pms interacted with normal matter/mis in the microcosm, an electromagnetic force was produced which caused normal matter/mis to fold in on itself. This increased available surface area needed to accommodate circuitry as normal matter/mis began configuring more complexly in reaction to the evolving complexity of relationship. Accordingly normal matter/mis convoluted with columns where 'minute circuits' that function as processors (perhaps with billions of components) are centered. Therefore intricate functions duplicate over and over the more convoluted normal matter/mis is configured.

Normal matter/mis is configured uniquely in the microcosm and accordingly functions differently. Although it is configured differently, what is similar is that its circuitry generates data patterned as electrical activity and creates electromagnetic radiation (EMR): waves of energy that travel at the speed of light in which electric and magnetic fields vary simultaneously. Similarly, patterns of vibrational activity generated by the open and closed strings of a FPDM/pms creates EMR. Thus patterns of electrical activity occurring in normal matter/mis and patterns of vibrational activity occurring in strings (i.e., open and closed) are both forms of EMR. The difference between them lies in their frequency, (i.e., how often waves of energy peak and trough in a given second).

Electrical patterns generated by the active circuitry of normal matter/mis are transmitted by way of EMR to FPDMs/pms. Therefore the vibrating open and closed strings of a FPDM/pms receive normal matter's data and they transform it by way of patterns of oscillation (i.e., vibrational activity). In a similar manner, vibrational patterns generated by open and closed strings in their excited state are transmitted to normal matter/mis by EMR. Therefore active circuitry of normal matter/mis receives the data of a FPDM/pms and they transform it by way of patterns of electrical activity. Accordingly FPDMs/pms and the normal matter/mis that embodies them are in a communicative structured relationship. Correspondingly they influence each other during every instant, but where there may be a delay in time between the moment the data is sent and the moment the information is received.

Cyclic (although not repetitive) relationship in the microcosm

Mind having expressed itself in a way that manifested three states (pas, pms and mis) in the macrocosm, expresses itself by way of innumerable relationships that take place in the microcosm. While in an embodied relationship in the microcosm, FPDMs/pms with primordial consciousness and normal matter/mis with the intelligence of conscious events each 'play' an integral part in their relationship. It is a two-way emergent process where their relationship evolves by way of unconscious intelligence (i.e., primordial consciousness) and the intelligence of conscious events. The relationship is then communicated as conscious intelligence (i.e., 'consciousness'):

- Unconscious intelligence: Due to the transformation of dark energy/pas to FPDMs/pms, the nature of mind is expressed as unconscious intelligence (i.e., primordial consciousness) that is intelligence of awareness and is dualistically transmitted by way of the vibrations of two types of strings (i.e., open and closed).

- The intelligence of conscious events: Due to normal energy transforming to normal matter, the unmanifest state of mind mental images manifests as the intelligence of conscious events. This intelligence is highly reactive and 'informative,' and provides a useful dimension to the relationship between FPDMs/pms and normal matter/mis by way of data.

- Conscious intelligence ('consciousness'): When FPDMs/pms and normal matter/mis first began embodied relationships in the microcosm billions of years ago, their relationship was a volatile two-way emergent process. Accordingly changes in their affective reactivity correlated with the degree of plasticity occurring in normal matter/mis with the intelligence of conscious events (12). Therefore the nature of mind expressed by embodied relationships that existed billions of years ago could be characterized as one of little acquired knowledge or understanding through thought.

Flexibility is a distinctive attribute of embodied relationships between FPDMs/pms and normal matter/mis. It describes a potential flow of that relationship as the degree by which it may be easily modified and thus is the predictor of change. Where no flexibility exists in relationships, cognition will not be expressed as part of conscious intelligence (i.e., 'consciousness') (13).

Vibrating strings, the 'record keepers' of relationships

Embodied FPDMs/pms with open strings and closed strings interact with normal matter/mis in a myriad of ways. Strings (i.e., open, closed) always vibrate in either their ground state (i.e., vibrating with a minimal amount of energy and the lowest mode of vibration) or their excited state (i.e., vibrating at an elevated energy level which is more than the ground state). While FPDMs/pms with primordial consciousness are embodied, strings change vibrations from their ground state to their excited state once enlisted. They are enlisted due to primordial events of unconscious intelligence (i.e., primordial consciousness). Enlisted strings (i.e., open, closed) vibrate in different patterns. This is due to their reactivity to signals they receive from normal matter/mis with the intelligence of conscious events. Strings attached to a C-brane are part of the configuration of FPDMs/pms that all share the same origin. All FPDMs/pms in the microcosm are derived from the transformation of dark energy/pas in the macrocosm and thus the former (i.e., dark energy/pas) is an extension of the latter (i.e., FPDMs/pms).

Accordingly the oscillations of two types of strings (i.e., open, closed) are the means by which *pure awareness* as represented by the dark energy of the macrocosm, manifests as *motion of awareness* (i.e., instead of pure awareness) in the microcosm. Although their exact number is unknown, each FPDM/pms interacting with normal matter/mis for the first time will have an equal number of closed strings and open strings which neither increases nor decreases:

- Enlisted open strings: vibrate in a way that correlates with their reactivity to the signals they receive from normal matter/mis that act as stimuli. Open strings (once enlisted) vibrate in a way whereby there is *no modification* as relates to their reactivity to stimuli (i.e., signals).

  Accordingly open strings vibrating in their excited state represent reactions of awareness by way of periodic motion (i.e., oscillation) as a response to the activity of the intelligence of conscious events. Therefore open strings consistently react with vibrations whereby normal matter/mis will be the state of mind that predominantly expresses the embodied relationship. This creates the data for the 'consciousness' of open strings and the nature of mind will be communicated by way of *habitual* behavior patterns.

- Enlisted closed strings: vibrate in a way that correlates with their reactivity to the signals they receive from normal matter/mis that act as stimuli. Closed strings (once enlisted), vibrate in a way whereby there is *a degree of modification* as relates to their reactivity to stimuli (i.e., signals).

Accordingly closed strings vibrating in their excited state represent reactions of awareness by way of oscillation (i.e., periodic motion) as a response to the activity of unconscious intelligence (i.e., primordial consciousness). Therefore closed strings consistently react with vibrations where FPDMs/pms will be the state of mind that predominantly expresses the embodied relationship. This creates the data for the 'consciousness' of closed strings and the nature of mind will be communicated by way of *'flexible'* behavior patterns.

If the relationship communicates the nature of mind by way of the 'consciousness' of open strings or the 'consciousness' of closed strings this causes string conversion. Accordingly strings are not destroyed but are only converted and thus an open string converts to a closed string or a closed string converts to an open string.

Enlisting vibrating strings, the 'record keepers' of relationships

Open and closed strings vibrating in their ground state receive the flow of data from normal matter/mis as signals transmitted by way of EMR. Accordingly a FPDM/pms, with unconscious intelligence and in the moment, is aware of the number of strings needed to accommodate the flow of data generated by normal matter/mis. Therefore a FPDM/pms engaged in primordial events (and thus aware of the population of its strings), enlists the number of strings (i.e., open, closed) needed on a momentary basis. This means that not every string of a FPDM/pms may be enlisted. There are some strings that may remain in their ground state and in this state (open, closed strings share the same ground state) one end of the string is attached to the C-brane and one end is open. When enlisted to transform data (generated by active circuitry as electrical activity) occurring in normal matter/mis, strings (i.e., open, closed) reactively change their vibrations to their exited state. Accordingly, strings (i.e., open, closed):

- vibrate similarly when in their ground state and while transforming data for sense impressions.
- vibrate differently when transforming data for mental events.

In this way, the vibrations of two types of strings (i.e., open and closed) attached to a C-brane represent the *potential* for dualistic expression of relationship. The ends of a string attached to a C-brane may move in one or more dimensions or the ends of the string may 'fix' to the C-brane. Enlisted open strings vibrate in a way that correlates with their reactivity to the signals they receive from normal matter/mis. Where there is no modification as relates to their reactivity to stimuli (i.e., signals), enlisted closed strings vibrate in a way that correlates with their reactivity to the signals they receive from normal matter/mis and where there is a degree of modification as relates to their reactivity to stimuli (i.e., signals).

Strings (i.e., open, closed) that reactively change their vibrations from their ground state to their excited state once enlisted, will vibrate similarly. They do so when receiving signals from normal matter/mis representing data to be transformed for sense impressions. Accordingly the C-brane functions similarly to a frictionless one dimensional 'object.' Although strings (i.e., open, closed) of FPDMs/pms when in their ground state have one string end attached to the C-brane, once enlisted (to transform data for sense impressions) both ends of a string may slide up and down (as if doing so within a 'frictionless hoop'). The curvature (i.e., gradient or differential) of the string, however, is constrained at its end points. Therefore the vibrations of open and closed strings while transforming the data of normal matter/mis for sense impressions, are similar.
Strings (i.e., open and closed) that reactively change their vibrations from their ground state to their excited state once enlisted, will vibrate differently. They do so when receiving signals from normal matter/mis representing data to be transformed for mental events:

- Enlisted open strings change their vibrations from their ground state to their excited state by bending so that their free end fixes to the C-brane. Thereby open strings now have fixed end-points with both ends attached to the C-brane. This specifies the conditions of the end-points and determines the frequencies at which open strings may vibrate.

- Enlisted closed strings change their vibrations from their ground state to their excited state by bending so that their free end 'taps in' to the C-brane. Thereby the free end of a closed string now penetrates the C-brane and whirls into compacted multi-dimensional dark energy/pas. Although the free end of a closed string may not be 'seen' whirling beneath the surface of a C-brane, it may whirl until it meets and connects with the other end of the string. In this way, the 'tapping in' and whirling by the free end of the closed string functions as a kind of momentum (i.e., movement of awareness). The 'tapping in' and connecting by the free end of the closed string vibrating in its excited state, however, is temporary and the free end eventually disconnects from the other end of the string. When this happens, the string recoils and configures to its previous ground state (one end attached to the surface of a C-brane and the other end free). Accordingly closed strings in their ground state receive signals from normal matter/mis in a way which is similar to open strings vibrating in their ground state.

Normal matter/mis generates data incessantly and FPDMs/pms will engage in primordial activity (i.e., primordial events) and create primordial algorithms for sense impressions and mental events. While in an embodied relationship, electrical patterns generated by the active circuitry of normal matter/mis are transmitted via EMR (i.e., electromagnetic radiation) to FPDMs/pms. Accordingly the open and closed strings of a FPDM/pms vibrating in their excited state transform the data of normal matter/mis by way of patterns of oscillation (i.e., vibrational activity). This creates a dualistic response to the relationship and as electromagnetic waves (EM waves), oscillating magnetic and electric fields. Consequently the vibrations of strings (i.e., open, closed) in their excited state create EM waves of open strings and EM waves of closed strings. The waves represent data of relationship between FPDMs/pms and normal matter/mis. This data will be encoded by a FPDM/pms engaged in primordial activity (i.e., primordial events) into primordial algorithms for sense impressions before it is encoded into mental events. Once primordial algorithms are created, they will be linked and the order and linkage of primordial algorithms will determine the content flow of 'consciousness' (i.e., conscious intelligence). This will communicate the nature of mind characterized by way of eight distinct attributes (to varying degrees): linguistic, logical-mathematical, musical, bodily-kinesthetic, spatial, interpersonal, intrapersonal, and naturalist (14). Accordingly, unconscious intelligence manifesting as primordial events (i.e., primordial activity) precedes the expression of relationship as conscious intelligence (i.e., 'consciousness') and where sense impression is communicated before mental event.

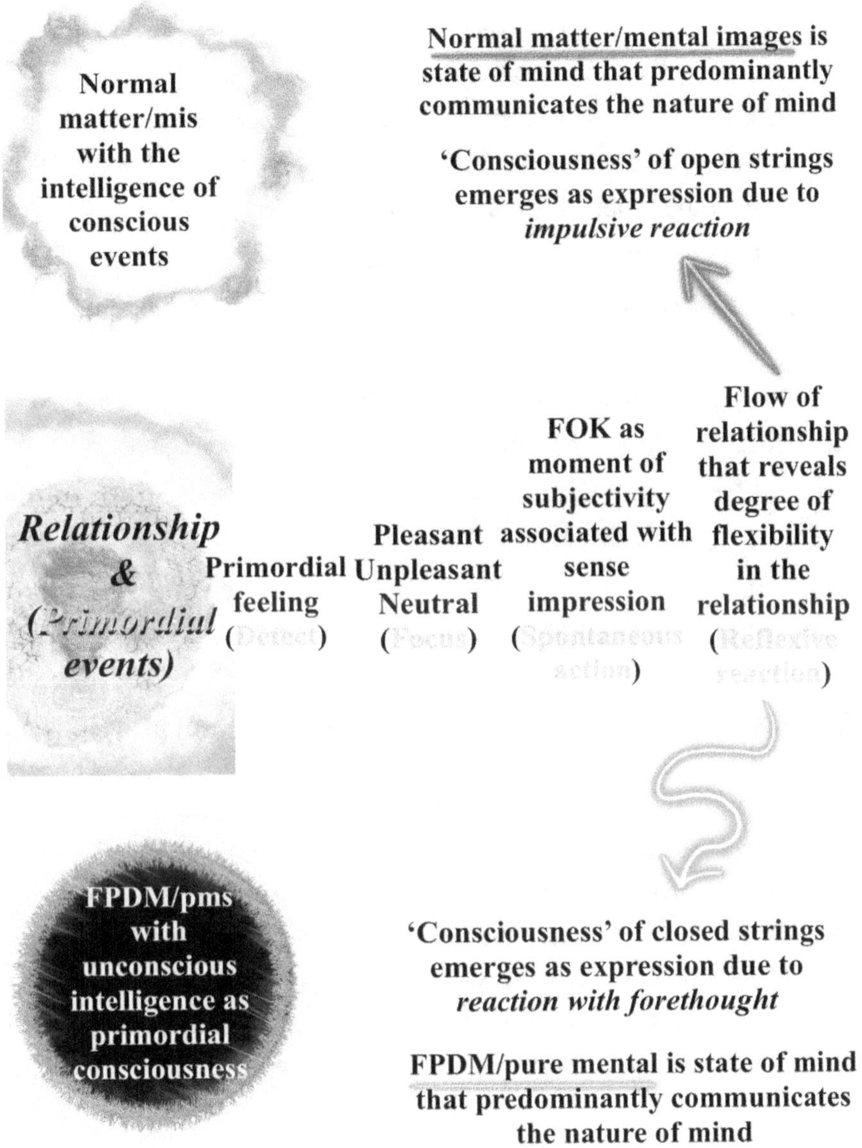

Figure 2

Primordial algorithms, rules of relationship (see Figure 2)

Sense impressions are sense organ perceptions (15):

- exteroceptive body (e.g., sight, hearing, touch, smell, taste, thermoception, pain)
- proprioceptive senses (e.g., position, motion state)
- interoceptive body (e.g., physical sensations)

Mental events and conscious percepts are specific content of conscious intelligence 'consciousness' (15) as emotions, thoughts, memories (16) and affective feelings, examples:

- perceptual stimuli (e.g., inner speech, dreams, visual imagery)
- fleeting present and its fading traces in immediate memory
- emotions (e.g., happiness, sadness, fear, anger, surprise, embarrassment, jealousy, guilt, and pride) (17)
- autobiographical episodes (experienced and recalled)
- expectations and effortful voluntary control
- explicit beliefs (about 'self', about the world)
- novel skills
- abstract concepts
- affective feelings (e.g., nociception, disgust, empathy)

FPDMs/pms with unconscious intelligence (i.e., primordial consciousness) while engaged in primordial events will express awareness with movement that will be both generalized and specific (see image 2). The movement of awareness will be generally expressed and thus 'radiant' when FPDMs/pms are methodically scanning patterns within EM waves. Alternatively the movement of awareness (due to stimulated emission) will be specific while FPDMs/pms *detect* patterns present in EM waves; this creates primordial feeling. They will then *focus* on the variations of patterns within EM waves whereby variation is grouped; this creates pleasant, unpleasant, or neutral Feelings of Knowing (FOKs) as events which act as stimuli. A FPDM/pms by way of spontaneous action will then capture with the movement of awareness small amounts of data (i.e., 'atoms').

These 'atoms' will be reflexively encoded into primordial algorithms for sense impressions which are melded with pleasant, unpleasant, or neutral FOKs. The moment that the relationship between a FPDM/pms and normal matter/mis is expressed as an 'object' of awareness in the form of a sense impression (i.e., pleasant, unpleasant, or neutral), the relationship spontaneously transforms. Accordingly it shifts from one that expresses unconscious intelligence (i.e., primordial events) and the intelligence of conscious events into a relationship that communicates 'consciousness.' However the expression of 'consciousness' is simply a pleasant, unpleasant, or neutral sense impression.

Due to the relationship between FPDMs/pms and normal matter/mis evolving over time, cognition is included as an attribute of conscious intelligence (i.e., 'consciousness'). The 'consciousness' of open strings, the 'consciousness' of closed strings, or a combination of both (i.e., consciousness) will predominantly express the relationship in a given moment. This begins with a momentary reaction to a sense impression (i.e., pleasant, unpleasant, or neutral). Accordingly an *impulsive reaction,* a *reaction with forethought,* a *reflexive reaction* to a sense impression, determines the manner by which the relationship gets expressed as an emergent property of 'consciousness.' Therefore a 'moment of subjectivity' to a sense impression manifests as an *impulsive reaction* (i.e., attachment or aversion) or is 'overridden' due to *reaction with forethought* (i.e., equanimity) or *reflexive reaction* (i.e., unnoticed). Accordingly the degree of flexibility in relationship is revealed by way of reaction and there is consequence.

Whatever reaction (i.e., impulsive reaction, reaction with forethought, reflexive reaction) to a sense impression (i.e., pleasant, unpleasant, or neutral) manifests acts as a stimulus. Accordingly a FPDM/pms will then reflexively encode 'atoms' into primordial algorithms for mental events (in a similar manner as it did when encoding algorithms for sense impressions). These are melded with the stimuli FOKs (i.e., pleasant, unpleasant, or neutral). Therefore primordial algorithms will be linked and this creates a flow of 'consciousness,' and content:

- *impulsive reaction* to a sense impression (i.e., pleasant, unpleasant, or neutral): normal matter/mis is the state of mind that predominantly communicates the nature of the relationship and by way of the 'consciousness' of open strings.
- *reaction with forethought* to a sense impression (i.e., pleasant, unpleasant, or neutral) implies that an impulsive reaction was 'overridden' and thus a FPDMs/pms is the state of mind that predominantly communicates the nature of the relationship and by way of the 'consciousness' of closed strings.
- *reflexive reaction* to a sense impression (i.e., pleasant, unpleasant, or neutral): directs primordial events whereby neither normal matter/mis nor a FPDM/pms is the state of mind that predominantly communicates the nature of the relationship. Instead it is consciousness that communicates the nature of the embodied relationship as a continuous stream of sense impressions and mental events. These arise and pass away and the nature of mind manifests in a way that would be categorized as being 'unnoticed.' Accordingly it is a FPDM/pms engaged in primordial events (i.e., primordial activity) with the movement of awareness as 'radiant' (i.e., generally expressed) and by way of 'stimulated emission' (i.e., expressed specifically) that overrides the dualistic movement of its vibrating strings (i.e., open, closed).

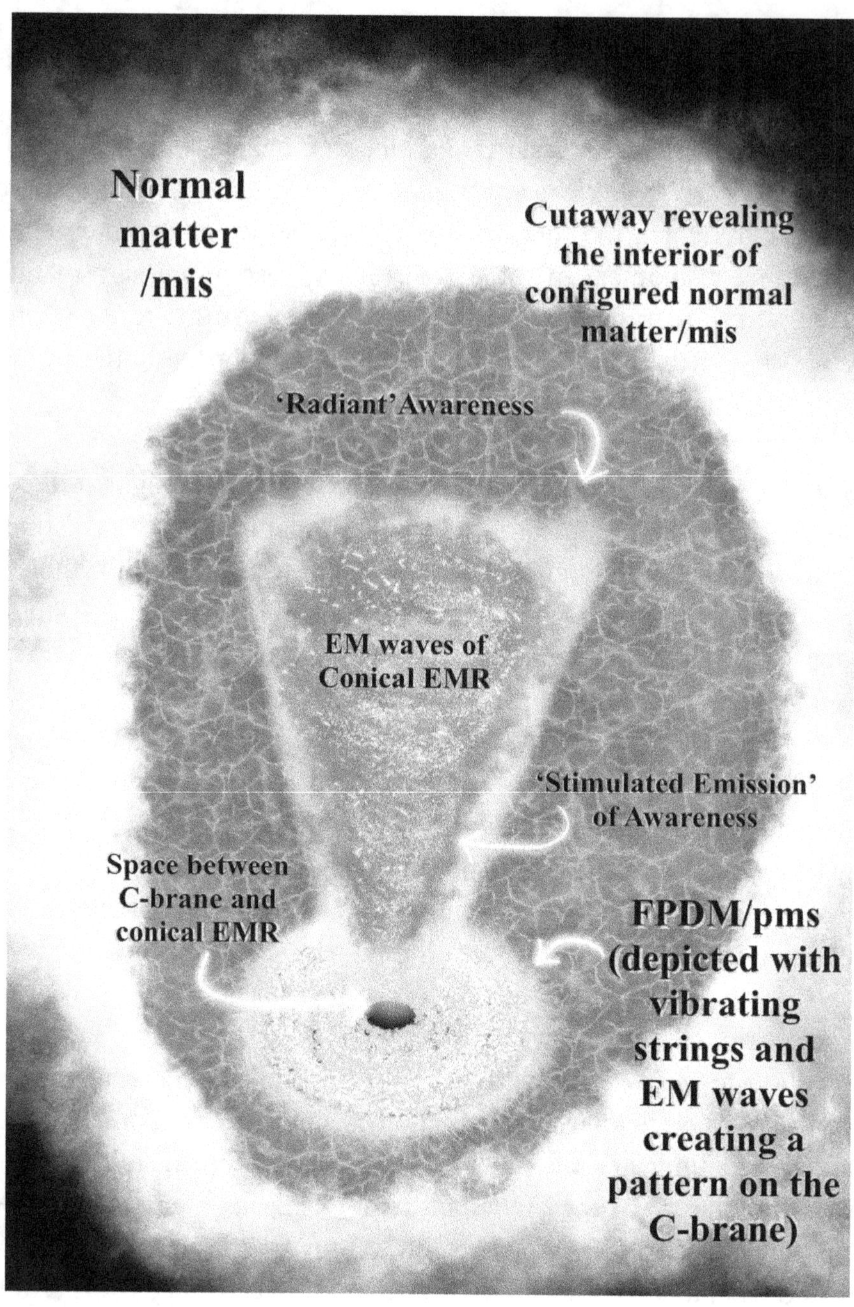

Figure 3

Conical EMR, shape of relationship (see Figure 3)

The vibrational activity of strings (i.e., open, closed) in their excited state transforms the data of normal matter/mis for sense impressions and mental events into electromagnetic radiation (EMR). Accordingly electromagnetic fields are created and transport the data of normal matter/mis as electromagnetic waves of energy (i.e., EM waves). These EM waves radiate onto the C-brane of a FPDM/pms and by doing so partially block the movement of awareness as a consequence of the relationship. Although gravity in the microcosm is too weak to configure normal matter/mental images that embodies FPDMs/pms, it 'pushes' the relationship. Accordingly gravity in the microcosm (in contrast to the macrocosm) is the movement of awareness that manifests as a 'push force' (18).   Thereby the force of gravity pushes EM waves towards the center of the C-brane. The EM waves that are pushed by a gravitational field bend and configure into conical EMR.

The emergent two-way interplay that flows between a FPDM/pms and normal matter/mis that embodies it continuously generates EM waves. These waves radiate frequently on the surface of the C-brane. Although strings (i.e., open, closed) are located on the same C-brane, those that transform data for mental events (once enlisted) form an outer layer of vibrating strings. Accordingly they generate more waves and at different frequencies than vibrating strings (i.e., open, closed) forming an inner layer that transforms data for sense impressions. This creates a functional pattern on the C-brane as a centralized shape of a sphere and, thus, a spheroidal region that is surrounded by EM waves.

Accordingly, present on the C-brane is a pattern resembling a sphere of revolution which represents the movement of awareness whereby EM waves pushed toward it configure into conical EMR (i.e., the shape of relationship). The base of the conical EMR contains the least amount of EM waves and represents the data of normal matter/mis transformed for sense impressions. As the EMR gets higher it widens due to more EM waves being present and represents the data of normal matter/mis transformed for mental events. The conical EMR acts as a space-filling field present within configured normal matter/mis. The way conical EMR radiates creates a region between the C-brane and the base of the EMR. The EM waves are diffuse material that represent a relationship with potential that is in orbital motion and thus are being pulled by the sphere of revolution as the motion of awareness (i.e., represented by a centralized spheroidal region). However, it has too much angular momentum and thus it momentarily escapes being pulled into the C-brane.

Due to the pattern present on the C-brane, the movement of awareness of a FPDM/pms with primordial consciousness is blocked in some directions by EM waves. However, awareness freely flows from the centralized spheroidal region. Therefore a FPDM/pms engaged in primordial events (i.e., primordial activity) of unconsciousness intelligence expresses awareness generally. Accordingly the movement of awareness is 'radiant' while a FPDM/pms methodically scans patterns within EM waves of conical EMR. Alternatively a FPDM/pms amplifies the movement of awareness by way of process, stimulated emission. This allows a FPDM/pms to use *fast mapping* to detect billions of patterns within EM waves while focusing on those which differ (19).

While engaged in spontaneous action, a FPDM/pms will capture 'atoms' (i.e., small amounts of data) and reflexively encode them into primordial algorithms for sense impressions. These are melded with FOKs (i.e., pleasant, unpleasant, or neutral). Therefore a 'moment of subjectivity' and reaction to a sense impression manifests as an *impulsive reaction* (i.e., attachment or aversion) or is 'overridden' due to *reaction with forethought* (i.e., equanimity) or *reflexive reaction* (i.e., unnoticed). Accordingly the relationship between a FPDM/pms and normal matter/mis transforms. This causes a FPDM/pms to reflexively encode 'atoms' for primordial algorithms for mental events that are melded with FOKs (i.e., pleasant, unpleasant, or neutral) and thus the nature of the relationship is communicated by way of conscious intelligence (i.e., 'consciousness').

A FPDM/pms will then engage in primordial events that correspond with complex warping of the C-brane and the centralized spheroidal region (as part of the pattern on the C-brane). Consequently the gravitational pull placed on the conical EMR (i.e., the shape of relationship) alters and thus the conical EMR is pulled into the C-brane (multi-dimensional space/awareness). The EMR distorts due to the action of dissolving into the C-brane and represents 'a transferred property' as fixed energy. This fixed energy propagates and represents awareness in motion as a consequence of the relationship that moves in the C-brane within an electromagnetic field. This transfers to strings (i.e., open, closed) attached to the C-brane and produces a fixed energy of EMR as photons that express a momentarily transformed relationship.

The 'moment of subjectivity' that manifested as attachment/aversion, equanimity or neither to a pleasant, unpleasant, or neutral sense impression determines the degree of string conversion. The strings convert in the amount that correlates with the degree of reaction (i.e., more reaction-more stings convert, less reaction-fewer strings convert):

- Where the expression was attachment/aversion and the reaction was impulsive, closed strings convert to open strings.
- Where the expression was equanimity and the reaction with forethought, open strings convert to closed strings.
- Where the reaction was reflexive, no strings convert.

Communicating relationship

The microcosm is a spatial structure containing innumerable unique configurations of normal matter/mis that, while embodying FPDMs/pms, forms a network. This network serves as the mechanism by which normal matter/mis relays data, although there may be limited assuredness as to the direction of information flow throughout this network. The way in which normal matter/mis is configured affects information flow as it travels throughout the network. A

FPDM/pms embodied by normal matter/mis shares information. The strings (i.e., open, closed) of a FPDM/pms vibrating in their excited state receive (by EMR) the data of normal matter/mis generated by active circuitry and transform it via patterns of vibrational activity. Similarly vibrational patterns generated by strings (i.e., open, closed) in their excited state are transmitted (by EMR) to the circuitry of normal matter/mis and transformed into electrical activity. When their communicative and shared relationship gets inserted as data and a set of instructions (i.e., code) transmitted along the network of normal matter/mis, it may influence 'other' embodied relationships. In this way, each embodied relationship that receives the data may serve as the means by which to 'execute it' by way of an expression of conscious intelligence (i.e., 'consciousness'). Accordingly a unique embodied relationship may replicate parts of its 'interplay' by modifying data traveling along the network of normal matter/mis.

The dark energy/pas of the macrocosm serves as the 'orchestrator' of communication between a pair or a group of embodied relationships. Because each FPDM/pms is derived from the same dark energy/pas, fundamentally, they truly cannot be described as having an independently existing relationship separate from 'other' FPDMs/pms (although a FPDM/pms is embodied by normal matter/mis as an individual). Accordingly communication between them represents the sum of infinite interdependent events as embodied relationships. Dark energy as the universe's pas represents existence to be unlike that of embodied FPDMs/pms. This results in a cause and effect relationship distinct in nature between the whole (i.e., the dark energy of the macrocosm) and its parts (i.e., FPDMs of the microcosm). Accordingly dark energy/pas having 'conceived ' FPDMs/pms forms an inextricable link between them. Therefore dark energy/pas facilitates communication between FPDMs/pms in a manner by which they act on one another without having any information relayed between them. Due to their being inextricable phenomena, communication between them is naturally simultaneous and such that one FPDM/pms *infers* the presence of the other (20). However each FPDM/pms and normal matter/mis brings distinctive characteristics to their relationship and thus will not have exactly the same properties as those of other embodied relationships.

Animate and inanimate

The relationship between a FPDM/pms and normal matter/mis is represented by way of an emergent property of conscious intelligence ('consciousness') and as a *vehicle of expression* in the form of a unique living being. The living being of conscious intelligence (i.e., 'consciousness') serves as a 'visual' indicator as to the kind of embodied relationship that exists between a FPDM/pms and normal matter/mis.

Correspondingly all 'objects,' including those in the form of living beings with conscious intelligence (i.e., 'consciousness'), represent relationships between the three states of mind (i.e., pure awareness, pure mental, mental images). However as relates to *animate* as opposed to *inanimate,* these terms denote different types of relationships between the states of mind. Animate 'objects' as *living beings* (e.g., invertebrates, mammals, birds, amphibians, reptiles and fish) represent *embodied relationships* between FPDMs/pms and normal matter/mis. These embodied relationships will be configured by way of the movement of vibrating open and closed strings (i.e., movement of awareness) that configure normal matter/mis. Inanimate 'objects' represent relationships between the three states but where they are not in an embodied relationship and include those that emerge as *living entities* (i.e., plants).

Embodied relationships and the Big Crunch

Each embodied relationship between a FPDM/pms and normal matter/mis has a unique timeline for when their interaction will end. Accordingly the number of embodied relationships in the microcosm increases or decreases according to the momentary occurrence where normal energy/mis is present but not expressed. When a particular relationship between a FPDM/pms and the normal matter/mis that embodies it ends, an unknown amount of normal matter/mis transforms into its previous form as normal energy/mis, present but not expressed. Due to the abundance of embodied relationships in the microcosm in which non-synchronized interactions begin and end normal energy/mis is in high demand. When normal energy/mis is unavailable, there will be FPDMs/pms that exist in the macrocosm demarcated from dark energy/pas by strings vibrating in their ground state but without a framework of normal matter/mis.

Due to strings converting rather than being destroyed, FPDMs/pms act as the 'record keeper' of individual embodied relationships. Accordingly a FPDM/pms will carry with it from one interaction to another a population of available open and closed strings attached to a C-brane. This represents a mechanism by which a continuity in embodied relationships is established and manifested as cause and effect. This includes the manner by which normal matter/mis gets configured. Because of the movement of awareness represented by the vibrations of open and closed strings, there will be transformation that advances expression (by way of continued relationship) or reverses expression (by way of discontinued relationship):

- The more open strings a FPDM/pms has available (due to closed strings converting to open strings), the lower the degree of complexity in the configuration of normal matter/mis is. Accordingly there is less potential for flexibility (to varying degrees) in the relationship and less potential for cognition to emerge as an attribute of the conscious intelligence ('consciousness'). Therefore the capacity to alter behavior as a result of experience is lower (21). This means the 'consciousness' of open strings emerges as the expression that communicates the relationship more so than the 'consciousness' of closed strings and is due to previous *impulsive reactions*. Therefore, in most cases (on a momentary basis), normal matter/mis is the state that predominantly communicates the nature of mind as an expression of this embodied relationship.

- The more closed strings a FPDM/pms has available (due to open strings converting to closed strings), the greater the degree of complexity in the configuration of normal matter/mis is. Accordingly there is more potential for flexibility (to varying degrees) in the relationship and more potential for cognition to emerge as an attribute of the conscious intelligence ('consciousness'). Therefore the capacity to alter behavior as a result of experience is greater (21). This means the 'consciousness' of closed strings emerges as the expression that communicates the relationship more so than the 'consciousness' of open strings and is due to previous *reactions with forethought*. Therefore, in most cases (on a momentary basis), FPDMs/pms is the state that predominantly communicates the nature of mind as an expression of this embodied relationship.

There are string populations of a FPDM/pms that represent the continuity of previous embodied relationships that result in 'maximal' consequence:

- When a FPDM/pms has a population of strings is which only closed strings are available except for one open string (due to open strings converting to closed strings), this may prevent a FPDM/pms from participating in perpetual interactions with normal matter/mis.

- When a FPDM/pms has a population of strings in which only open strings are available except for one closed string (due to closed strings converting to open strings), this may prevent a FPDM/pms from participating in perpetual interactions with normal matter/mis.

Accordingly the strings (i.e., one open and the rest closed, or one closed and the rest open) which demarcated a FPDM/pms from the universe's dark energy/pas may dissolve into a C-brane. This adjusts initial conditions in the macrocosm by a FPDM/pms.

Alternatively this string population may cause interactions with normal energy/mis, when available, to be pulled by the consequence of gravity. Accordingly normal energy/mis (unmanifest) will configure as normal matter/mis (manifest) with the intelligence conscious events. Therefore a FPDM/pms in an embodied relationship with normal matter/mis with continuity from their past (due to the population of open and closed strings of a FPDM/pms) will communicate with 'other' FPDMs/pms and normal matter/mis in embodied relationships. They will do so however, with either a lesser or greater degree of flexibility, which will have consequences as relates to the 'other' structured relationships present in the microcosm.

What seems to be a cause and effect relationship may only be possible if neither the cause nor the effect exists independently and permanently (3). In this way, and as relates to the scenarios presented, the outcome is uncertain (i.e., not accessible by way of cognition). This is because the outcome is dependent on the movement of pure awareness. Accordingly dark energy is the facilitator of the outcomes, rather than the movement of awareness due to the vibrations of strings (i.e., open and closed).

The Big Bang(s) may represent the start of this cycle which is one of an infinite number of cycles of interactions between FPDMs/pms and normal matter/mis. The cycles may stop, as a consequence of gravity. This represents the movement of dark energy/pas at the macrocosm. Accordingly instead of universal expansion there may be a Big Crunch and thus transformation of the macrocosm which has collapsed into itself. Therefore the macrocosm will be absorbed into isotopic space, whereby a new cycle of mind expressing itself differently may manifest (3).

Conclusion

This theory melds features from different theories with the aim of assisting understanding of mind and the expressions of mind rather than trying to dogmatically prove the 'correctness' of any one theory. Nevertheless no theory may ever account for the state of mind pure awareness which exists entirely separate from 'consciousness.' Admittedly the state of pure awareness may not be accessible to thought but this may not prevent an individual from visualizing this state as being represented by a particular phenomenon (i.e., dark energy). An individual may associate 'objects' (i.e., sense impressions and mental events) with *everyday existence* which may be perceived similarly or dissimilarly by individuals.

Accordingly 'objects,' when arranged in a particular manner may construct a viewpoint that allows an individual to pick out a particular feature and emphasize it. The act of doing this, however, may create understanding or confusion as relates to reality (i.e., expressions of mind). Phenomena when 'cognitively linked' may formulate a theory and, when made on the basis of limited 'objects,' a hypothesis that when perceived similarly by many individuals is termed 'empirical' evidence. Accordingly theories, hypotheses or 'empirical' evidence may represent a myriad of unique viewpoints. Nevertheless what may be similar between them is their ability to 'highlight' a particular aspect(s) as relates to the relationship between the three states of mind.

Although mind may be unfathomable while thinking, the nature of mind may be experienced by the events of 'consciousness' as FOKs and an 'observing ego' and with perception of a living being and thus represent the relationships between the states of mind (i.e., the pure awareness state [pas]/the pure mental state [pms] along with the pure mental state [pms]/the mental images state [mis]). Accordingly mind may be all there is and all that will be with the potential to express itself differently.

Received June 29, 2020; Accepted July 25, 2020

References

(1) Pollard-Wright, H. (2020). Electrochemical energy, primordial feelings and feelings of knowing (FOK): Mindfulness-based intervention for interoceptive experience related to phobic and anxiety disorders. *Medical Hypotheses*, *144*, 109909. https://doi.org/10.1016/j.mehy.2020.109909

(2) Baumann, D. (n.d.). *"Cosmology: Part III Mathematical Tripos, Cambridge University."* [online] Available at: http://www.damtp.cam.ac.uk/user/db275/Cosmology/Lectures.pdf [Accessed 29 May 2020].

(3) Matthieu Ricard, Xuan Thuan Trinh and International Society For Science And Religion (2007). *The quantum and the lotus : a journey to the frontiers where science and Buddhism meet.* Cambridge: International Society For Science And Religion.

(4) Katrin Becker, Becker, M. and Schwarz, J.H. (2011). String theory and M-theory : A modern introduction = 弦论和M理论导论 / String theory and M-theory : A modern introduction = xian lun he M li lun dao lun. Beijing: World Book Publishing Company.

(5) Moore, G. (2005). WHAT IS… a Brane? *Notices of the American Mathematical Society*, 52, pp. 214–215.

(6) Khyentse, D. The Collected Works of Dilgo Khyentse: Volume One. Shambhala Publications. Boulder, CO. 2010.

(7) Klebanov, I.R. and Maldacena, J.M. (2009). Solving quantum field theories via curved spacetimes. *Physics Today*, 62(1), pp.28–33.

(8) Kibble, T. (2009). Englert-Brout-Higgs-Guralnik-Hagen-Kibble mechanism (history). *Scholarpedia*, 4(1).

(9) McMullin, E. (2002). Physics in Perspective, 4(1), pp.13–39.

(10) Silberstein, M., William Mark Stuckey, & Mcdevitt, T. (2018). Beyond the dynamical universe: unifying block universe physics and time as experienced. Oxford University Press.

(11) Budday, S., Raybaud, C. & Kuhl, E. A mechanical model predicts morphological abnormalities in the developing human brain. *Sci Rep* **4,** 5644 (2015). https://doi.org/10.1038/srep05644

(12) Ganguly, K., & Poo, M. Activity-Dependent Neural Plasticity from Bench to Bedside. Neuron. 2013. 80(3), 729–741.

(13) MacLean EL, Hare B, Nunn CL, et al. The evolution of self-control. *Proc Natl Acad Sci U S A.* 2014;111(20):E2140- E2148. doi:10.1073/pnas.1323533111

(14) Gardner, H. (1993). *Multiple intelligences : The Theory In Practice.* Basic Books.

(15) Baars, B. (2015). Consciousness. *Scholarpedia*, 10(8), p.2207.

(16) Grabovac, A. D., Lau, M. A., & Willett, B. R. (2011). Mechanisms of Mindfulness: A Buddhist Psychological Model. *Mindfulness*, 2(3), 154–166. https://doi.org/10.1007/s12671-011-0054-5

(17) Damasio, A.R. (2000). The feeling of what happens : body, emotion and the making of consciousness. London: Vintage, , Cop.

(18) Rancourt, L., & Tattersall, P. J. (2015). Further Experiments Demonstrating the Effect of Light on Gravitation. *Applied Physics Research*, 7(4). https://doi.org/10.5539/apr.v7n4p4

(19) Vlach, H. A., & Sandhofer, C. M. (2012). Fast Mapping Across Time: Memory Processes Support Children's Retention of Learned Words. *Frontiers in Psychology*, 3. https://doi.org/10.3389/fpsyg.2012.00046(19)

(20) Zieffler, A., Garfield, J., Delmas, R., & Reading, C. (n.d.). *A FRAMEWORK TO SUPPORT RESEARCH ON INFORMAL INFERENTIAL REASONING 5.* Retrieved June 21, 2020, from http://www.stat.auckland.ac.nz/~iase/serj/SERJ7(2)_Zieffler.pdf

(21) Eelen, P., Hermans, D., & Baeyens, F. (2001). Learning perspectives on anxiety disorders. In E. J. L., C., D., & J. (Eds.), Anxiety disorders : An introduction to clinical management and research (pp. 249–264). New York: Wiley.

Perspective

Interactionism, Evolution, & the Initial Alteration

Francis Schneider[*]

Abstract

Unlike many other articles in philosophy of the mind, this is not an article which aims to prove or even give evidence for an ontological position regarding the mind. This article *assumes* a position regarding the mind and then draws implications from it. This article aims to show that if we assume two things as given, that of an interactionist view point and the theory of evolution by mutation and natural selection, it follows that there is a particular moment in time I label the *initial alteration*. The initial alteration is the original instance of the change in behavior of matter in the brain of some organism in our evolutionary history. I will show that this moment in time is of great interest and opens the possibility of moving the mind-body problem towards experimentation. I list several possibilities surrounding this moment, though I will not explore them all in detail.

Keywords: Philosophy, mind, interactionism, evolution, initial alteration.

The Initial Alteration

In philosophy of the mind, there are a multitude of different positions one can take regarding the ontological nature of the mind in relation to the physical brain. Famously, the issue was broken into two stances by the philosopher Rene Descartes. The positions, Monism and Dualism, have been continued to be debated for hundreds of years. These two positions should be very well known to those interested in the mind-body problem. For the purposes of this article I am only going to give very basic definitions of each because comparing them is not my purpose here. I will also list two of their more popular sub-positions even though there are more.

A simplistic definition of Monism is that it attributes a kind of oneness to the mind and brain. Monism has positions such as Property Monism and Substance Monism. Property Monism is loosely the idea that all properties are a single type, and so all properties are physical properties - *or* they are all mental properties. Substance Monism is the view that only a single type of substance exists, and everything is some form or another of this substance, including the mind and brain.

Dualism believes that the mind and the brain are distinct and separable. Dualism has positions such as Interactionism and Epiphenomenalism. Interactionism posits that the mind and body are two separate entities which causally affect one another. Epiphenomenalism posits that mind and body are two separate entities, but the only causal affect is from the body on the mind, the mind does not causally affect the body. In this article, we are going to be looking at a sub-position for dualism. That of interactionism. Even though we are going to be assuming interactionism and

[*] Correspondence author: Francis Schneider, Independent Researcher. E-mail: francisjschneider@gmail.com

drawing implications from it, I believe many of my conclusions will be of importance to anyone interested in the mind-body problem.

A more formal definition of interactionism is given by the Stanford Philosophical Dictionary as (Robinson, 2017):

> [T]he view that mind and body—or mental events and physical events—causally influence each other". (Robinson, 2017) The specific part of interactionism I am going to be addressing is that the mind having a causal influence over the brain implies that some matter in the brain of humans (and possibly some other organisms) is behaving differently than would be expected if that matter had been governed completely by the conventions established for their behavior by physics and chemistry. This feature of the interactionist position draws opposition from some who say it seems to violate what is called the *physical closure* of physics – defined as "every physical event has a physical explanation.

Sometimes the term *causal closure* is used instead.

What is also important to understand for this article is the theory of evolution by mutation and natural selection. Only a rudimentary understanding is necessary, and I will give a brief overview for those who are not familiar.

When organisms reproduce (replicate their genetic codes), they create offspring. The genetic codes of their offspring are copies of their own code, but there are what are called *mutations* which are errors in the copying. The effect these mutations can have varies. They can be harmful, essentially ineffective, or beneficial.

Organisms who have beneficial mutations tend to have more offspring and their genetic codes (mutation included) get passed on to future generations. As this process happens over and over, organisms begin to change in a way that correlates strongly with them being better adapted to surviving in their respective environments. This process is the widely accepted scientific view on the development of life on earth. As stated, we are going to be talking about interactionism and its implied behavior changes on matter in the context of evolution.

A constant theme in the discussions surrounding the mind and its relation to the brain are what are called *The Neural Correlates of Consciousness* or "the minimum neuronal mechanisms jointly sufficient for any one specific conscious experience." (Wu, 2018) The Neural Correlates of Consciousness reference the set of objects in the brain which give rise to consciousness but does not exactly mention the reaction of the brain to consciousness.

Science has made progress through experimentation in identifying regions in the brain which seem to be strongly correlated with aspects of consciousness. This is a hugely important endeavor, but for the implications I am going to draw using interactionism and evolution, it is not important what *particular* objects are involved or at what scale the changes are taking place (i.e. the quantum probabilistic level or the more macro level of variation in the charges within neurons). It is only important that *some* objects in the brain at some level are causally associated with the mind in the way interactionism posits.

In the interactionist view, how does the mind evolve over time?:

For an example, we can look back to our nearest common ancestors with chimpanzees. This family of organisms is called *Homonini* by taxonomists. Assuming that chimpanzees have a more primitive mind and have some form of subjective experience such as experiencing sensations of pain and pleasure – which does seem to be the view held by the large majority of interactionists - it is safe to assume this creature also had a primitive mind and experienced *Qualia* (the name given to single instances of subjective experience). So how did we get from these organisms to humans today?

Like with any organism, there were mutations that happened when these organisms reproduced. One place these mutations happened was in the brain. Some of these mutations would have affected the *Neural Correlates of Consciousness* for that creature in some way. Some of these changes would have been harmful - perhaps rendering the creature mentally unstable. Others of these changes would have been beneficial – possibly making the creature more creative. The organisms who gained evolutionary benefit from the altering of their minds would have outcompeted their rivals and passed these genes onto the next generation.

This process happened repeatedly: a mutation leading to a change in the Neural Correlates of Consciousness for that creature (when compared to its parent), a change in behavior of matter in the brain, and the changes that lead to evolutionary benefit being carried over - until we arrive at humans today. Again, I will remind my reader that this follows if we accept both an interactionist perspective as well as accept the evolutionary model of the development of life. This would not necessarily apply to non-interactionist positions.

This is actually very similar to how *any* feature of an organism evolves except that most evolutionary changes that take place for an organism do not change the behavior of the matter itself, only utilize that matter in a way consistent with the conventions of physics and chemistry.

So in analyzing the evolution of the mind from an interactionist view we can see there is a process going on in the brains of some organisms where the behavior of matter is constantly being changed overtime via mutation and selection. Since life has only been around on earth for a finite time, this process must have had a beginning. Meaning, there was a first time the behavior of matter in the brain of some organism in our evolutionary past was altered from its behavior that would be expected if it were behaving purely by the conventions of physics and chemistry. So for those who believe in the *causal closure* of physics, this would be the initial instance of the causal closure being violated.

From an interactionist standpoint, the only alternative to this conclusion is that the initial alteration in the behavior of matter happened before organisms developed brains. This is an interesting possibility, especially from an experimental standpoint, but I will not be addressing it in this article. Here I will assume the initial alteration took place in the brain of some organism in our evolutionary past.

This initial instance of change has several questions surrounding it. Some questions that could be asked are: In what organism did this take place? How long ago in our evolutionary history did it take place? What objects in the brain were involved? How many objects in the brain were involved?

A question I want to highlight surrounding the initial alteration (this could also apply to those alterations that evolutionarily follow relatively soon afterword). *Was consciousness involved or not?* Or is it possible there was something else involved that consciousness later evolved from? The first idea seems like it might be appealing because the thing associated causally with the change in behavior of matter is something that we are familiar with: that of the mind or one of its features (such as a Qualia).

The other idea has been discussed, though. Bertrand Russell wrote about what he called *protophenomenal* properties. David Chalmers gives a good summery of Russell's view: "…protophenomenal properties are special properties that are not phenomenal (there is nothing it is like to have a single protophenomenal property) but that can collectively constitute phenomenal properties, perhaps when arranged in the right structure." (Chalmers, 2013) Bertrand Russell himself was not an interactionist as interactionism is part of dualism and Russell was a monist. That being said, I believe the basic essence of the idea can be translated over to an interactionist perspective. Specifically that consciousness arose from a smaller scale phenomenon, but here we would be positing that this phenomenon causally affected matter in the brain in a way that changed its behavior from that if it had been governed by the conventions of physics and chemistry. When we ask the question *was consciousness involved or not,* one could argue that such a view is more compatible with the smaller scale an *initial* change in behavior might seem to entail.

To look at one last possibility somewhat related to the previous one I described, I would like to examine mind-brain interaction in humans again. To do so, I will ask a question. *How many objects in the brain give rise to consciousness?* For the purposes here, a specific number is not needed. All that we need to observe is that more than one or two objects are involved in giving rise to and (if interactionism is true) reacting to the mind. The alternative to this: that our entire minds are being produced and reacted to by one or two objects, would seem absurd.

The fact that there are multiple objects that are causally associated with each other in a way different from the normal limitations of chemical reactions is a very important implication of the interactionist viewpoint. It is not the purpose of this article to examine this implication of interactionism, but I will point it out because I believe objects becoming causally associated with each other in ways they could not before could be a lead for examining the initial alteration both experimentally and logically, due to the possible evolutionary benefit this type of effect could produce even at a low-scale.

Experimentation

I believe the most important aspect about the initial alteration from the perspective of people interested in the mind-body problem today is the possibility of experimentation. The smaller scale changes that the initial alteration would seem to imply might be easier to detect and understand than the behavior changes of matter going on in the brains of humans.

If the view that consciousness arose from some smaller scale phenomenon is correct, (such as the interactionist adaption of the *protophenomenal properties* view), then we may be able to detect such a thing by replicating smaller-scale nervous systems involving matter thought to be used in

(Published in Journal of Consciousness Exploration & Research| August 2020 | Volume 11 | Issue 5 | pp. 536-540)
Schneider, F., *Interactionism, Evolution, & the Initial Alteration*

brains of organisms in the past. These types of experiments would be difficult but certainly possible with today's technology and dedicated people. If that view did not turn out to be true, perhaps we would have to experiment with a system more complicated before changes were detected. Even then, it might very well still be possible though more technological innovation would be required to produce the systems where the effect took place. Perhaps observing the brains of smaller organisms such as insects could provide inspiration.

Conclusion

I believe the *initial alteration* is an inescapable implication if we assume both interactionism and evolution by mutation and natural selection. For people who hold this view, many important questions arise from examining the issue. It also may provide substantial possibilities in the way of moving the mind-body problem further into the realm of experimentation.

Received June 6, 2020; Accepted July 25, 2020

References

Robinson, H., *Dualism*, The Stanford Encyclopedia of Philosophy (Fall 2017 Edition), Edward N. Zalta (ed.), https://plato.stanford.edu/archives/fall2017/entries/dualism/

Wu, W., *The Neuroscience of Consciousness*, The Stanford Encyclopedia of Philosophy (Winter 2018 Edition), Edward N. Zalta (ed.), https://plato.stanford.edu/archives/win2018/entries/consciousness-neuroscience/

Chalmers, D. J., *Panpsychism and Panprotopsychism*, The Amherst Lecture in Philosophy 8 (2013): 1–35, http://www.amherstlecture.org/chalmers2013/

ISSN: 2153-8212   Journal of Consciousness Exploration & Research   www.JCER.com
Published by QuantumDream, Inc.

(Published in Journal of Consciousness Exploration & Research| August 2020 | Volume 11 | Issue 5 | pp. 541-547) 104
Cocchi, M., & Gabrielli, F., *Brain, Mind & Soul: The Hard Problem of the Measurement*

Dialogue

Brain, Mind & Soul: The Hard Problem of the Measurement

Massimo Cocchi* & Fabio Gabrielli

Research Institute for Quantitative & Quantum Dynamics of Living Organisms,
Center for Medicine, Mathematics & Philosophy Studies, Italy

Abstract

The aim of the paper, in form of a dialogue, is to discuss the problem of the fracture between the reduction of brain phenomena to quantitative measure and the quality of individual feeling. If mathematical models that could make the unrepeatable forms of individual feeling measurable, therefore public, capable of going further the simple neuroanatomical mapping, then a huge problem of individual and public ethics would arise. In other words, there might be a risk of full control on our lives, on our intimacy: Foucaultian biopolitics and bio power could become even more pervasive.

Keywords: Brain, mind, soul, hard problem, measurement, dialogue.

Massimo

I still hardly understand how I joined in with a group of famous scientists (Cocchi 2019; Tarlaci 2019) that deal with quantum theory from different points of view, from physics to mathematics, from chemistry to medicine, as well as philosophy, etc.

Maybe I joined it because unpredictably one day, with a function magically drawn by a friend of mine, the mathematician Lucio Tonello, and masterfully explained by another friend of mine, the philosopher Fabio Gabrielli, I classified the most impenetrable moment of the human being, *depression* (Cocchi 2015; Tonello 2015; Cocchi 2014; Cocchi 2014; Cocchi 2015; Tonello 2015; Cocchi 2006).

Well, maybe this was my fatal mistake, the one that pushed me to discuss in an extremely difficult way with friends much more illustrious and famous than I am on the quantum subject, a matter of which I hardly understand the definition.
If the classical physics represents the measurable, the quantum physics represents the *classic* immeasurable.

---

*Correspondence: Prof. Massimo Cocchi, Research Institute for Quantitative & Quantum Dynamics of Living Organisms, Center for Medicine, Mathematics & Philosophy Studies; & Department of Veterinary Medical Sciences, University of Bologna, Italy.
E-mail: massimo.cocchi@unibo.it

So, what do measurable and immeasurable mean with regard to soul and mind? We can not consider the quantum aspects of the brain, the quantum brain, without considering its symbiosis with the brain classically understood. They coexist in one another and they realize mutually feeding in measurable and not measurable paths, or not yet potentially measurable. We think of measurable actions like the bodily gesture in all its expressions, and not measurable or not yet potentially measurable like the sphere of feelings, which refer to a series of neuro-related that avoid the feasible mathematization of biological-molecular pathways as we know them today.

I am a firm advocate of biomolecular dynamics of the animal organism (Cocchi 2017) and I believe, as we have gathered evidence with some friends at the Atomic Institute of Vienna (International Agency for Atomic Energy, Vienna) (Summhammer, 2020), that what we know about the cellular bio molecularity (interactome) to the ion channels is the *incipit* of so-called not measurable phenomena.

Not measurable because we still don't know them and to date we don't know if it will ever be possible to identify them.

Certainly, with the word "quantum" many famous physics and mathematicians have freed their imagination and, with some complicated equations, they have tried to demonstrate what is not perceivable.

Can we truly think that quantum physics by itself, without the classic one, will explain the dynamics of the universe?

In addition, is it possible that the brain, in what we know and what we don't know, follows the same dynamics that the universe follows? Indeed, is it possible that the brain is the basic unit of the universe with the same complexity, a microcosm, between classical mechanics and quantum mechanics, measurable, potentially measurable and immeasurable, know and not known?

Certainly, I made a series of banal observations, which my *quantum* friends would consider as a scientific blasphemy, however, we have captured the long shadow on the soul, the depression, the dark evil that has exalted the psychiatric hypothesis so much, that has seen so many words spent in an effort to explain the origin of this shadow (Cocchi 2010; Cocchi 2010).

When, in the relation among three numbers, we can identify pathological characteristics of the brain, intended as mind and behaviour, our thoughts turn to the concept of soul.

Will soul and mind ever be explained by reason?

Will we ever know where the soul sprouts along with its expressions and where the mind arises along with its expressions?

Could they be the quantum synthesis of the essence of the animal man?
Without soul and mind the animal man wouldn't *create* good or evil, he wouldn't be realized in the muddle of biochemistry and quantum physics.

How embarrassing is it to pronounce the words *soul* and *mind* without knowing exactly where they arise and what they are?

I remember that Sir John Eccles, discoverer of neurochemical and neuroelectric transmission, felt an urgent need to resort to the reflections of Karl Popper, certainly Sir John Eccles had discovered how information flows through the meanders of the brain, but he wasn't able to define the path up to the expression of soul and mind, and he was deeply troubled with that.
Philosophy is kind of like our soul and mind, it grows in brains that don't want to find the biochemical paths of soul and mind, but, giving up the discussion on measurable and not measurable, they try to interpret the meaning in the light of the human action.
At this point, after my questionable *incipit*, I'm forced to give the floor to Fabio Gabrielli, because he has to comfort me on my physical-mathematical ignorance giving me the illusion that a time will come when perhaps I will understand the molecular path of soul and mind, so far I need comfort in the face of my inability to understand the animal man (Cocchi 2017; Cocchi 2017).

Fabio

Anima (soul, spirit) is a bright Latin word that indicates the blow, the spirit, the vital principle; *animus* refers to the thinking spirit.

The Greek *psyché*, in turn, indicates the soul, the vital blow, the breath. Eraclito says that the *logos* of the soul (its measure) is too deep to catch it (fragment 45).

Anima, therefore, refers to the deep core of our being, to our intimacy never completely measurable, ascribable at calculation. According to Socrates and his heritage, it is the location of intellectual and moral qualities of the human being, his consciousness.

Anima is an extremely refined cultural construct, of relevant spiritual significance and extraordinary anthropological charm, with a range through different historical epochs. Philosophy, art and literature have brought a great deal of attention to the concept of *anima*, with particular reference to the relationship between soul and body.

In this respect, cognitive science and neuroscience provide relevant theoretical positions, promising hypothesis, fruitful theoretical constructs right on the crucial issue of the relationship between mind (*anima*) and body.

For instance, John R. Searle (Searle 1992) believes that consciousness is the distinctive feature of our mental life: in this sense, it is characterised by intentionality, that is to say, ability to refer to other than the self.

Besides, consciousness is an emerging biological property of the brain, for which indicates a systemic relationship, that is, qualitative among parts, not ascribable to its basic materials.

Daniel C. Dennett (Dennett 1992; Dennett 1987), conversely, maintains a reductionist functionalism, for which the mind doesn't represent an autonomous or privileged level of being. From a neurophysiological point of view, there are no mental states, but architectures, neuronal configurations, cortical starting-up models that lack intentionality.

David Chalmers (Chalmers 1996), with whom I mostly agree, maintains that consciousness represents an unsolvable *hard problem* if we trace it back into merely neurophysiological and functionalist explanations. Consciousness, mind/brain are two properties that refer to a common structure: overlapping information states, which are expressed on the basis of quantum mechanics. Everything is information processing with different degrees of complexity (Accorsi 2017; Cocchi 2017): even a toaster, a thermostat, a vase, a fridge can have minimal forms of consciousness (Tononi 2016). The ancient panpsychism returns in a very refined form. Martin Heidegger himself has continued to conceive the human animal as something separated from animality, nature, in the sign of the primacy of language (Heidegger 2010). Yet art, for example, can be the way to interrupt the myth of self-centered and hyper-measuring *homo sapiens*: a work of art can stay alive even without the presence of its artist; an artistic photograph, for example, lives on its world, it bears witness to the world, in the absence of the photographer. In short, "only the inhuman is photogenic" (Baudrillard 1992).

Lastly, I would like to mention briefly the *embodiment theory*, of corporeality of concepts, if you forgive my brutal synthesis, which has also been taken up by some very interesting theories about personal identity and self-awareness. In practice, the experience we have of ourselves is shaped by our sense of bodily self-awareness. Suffice is to think, as simple mention, about the primacy of the body plan of Vittorio Gallese (Gallese 2010; Gallese 1998) or about the fundamental role of the body in operational dynamics (in acting) and in the genesis of emotions of Antonio Damasio (Damasio, 2010).

Personally, I believe that consciousness is not reduced to mere neuroanatomy or neurophysiology, not even to reductionist functionalism. As some interesting attempts to provide a quantum measurement of consciousness, a sort of new state of matter where atoms process information from which the so called conscious subjectivity originates (Tegmark 2015), leave several aporias open, especially in relation to the quality of our conscious experiences.

The problem of the jump, of the fracture between the reduction of brain phenomena to quantitative measure and the quality of individual feeling can't be filled with the claim of an all-inclusive comprehensive organic math (quantitative reductionism). If one day there were mathematical models that could make the unrepeatable forms of individual feeling measurable, therefore public, capable of going further the simple neuroanatomical mapping, then a huge problem of individual and public ethics would arise. In other words, there might be a risk of full control on our lives, on our intimacy: Foucaultian bio politics and bio power could become even more pervasive. The path of science will absolutely have to take these two aspects into consideration: the possibility to achieve the intimacy of individual consciousness and the ethics of responsibility towards the exposed consciousness itself. Mathematical models, quantum physical approaches, experimental data that we will be able to produce won't merely highlight biological emergentism of consciousness, but also its ethical nature, the fact that the whole of its functions expresses a unique, singular, irreplaceable life, at any level of reality, therefore innumerable.

In other words, I think that a reductionist view of the world in the sign of matter (materialistic, mechanistic explanation of nature) has some limits, at the same time there is an autonomy of the mental that is irreducible to informatic dynamics of artificial intelligence (Searle 1997).

In conclusion, biology can't be a merely physical science, the gap between subjective and objective is too great for us to think about the mental only on the basis of our way of conceiving events and physical processes (Nagel 2012).

Massimo

I understood that my far run-up to explain the molecular dynamics of the long shadow on the soul is at a point of no return, but I have to get out of ideological beliefs that seemed to me essential. Too hastily I believed that identifying a psychopathology by measurable mechanisms could explain to me the concepts of soul and consciousness. Too hastily I believed that the whole thing could resolve in an essential and difficult molecular journey: on the contrary, I had reached a road sign whose arrow indicated a path still unknown, for which I don't have a compass showing me the way to go.

I take comfort in the fact that the possibility of recognising different psychopathological aspects can prevent the invasion of the brain with inappropriate medications from modifying that path, which is unknown for us, but certainly well defined by *creation*, which we temporarily call *soul* and *consciousness*.

(Published in Journal of Consciousness Exploration & Research| August 2020 | Volume 11 | Issue 5 | pp. 541-547)   109
Cocchi, M., & Gabrielli, F., *Brain, Mind & Soul: The Hard Problem of the Measurement*

Received June 21, 2020; Accepted July 25, 2020

References

Accorsi, P.A, Mondo, E., Cocchi, M. (2017). Did you know that your animals have consciousness? Journal of Integrative Neuroscience 16, S1–S2.

Baudrillard, J. (1992). The Disappearance of Art and Politics. Palgrave Macmillan, New York.

Chalmers, D. (1996). The Conscious Mind: In Search of a Fundamental Theory. Oxford University Press.

Cocchi, M., Tonello, L. (2010). "Bio molecular considerations in Major Depression and Ischemic Cardiovascular Disease". Central Nervous System Agents in Medicinal Chemistry 9: 2-11.

Cocchi, M., Bernroider, G., Rasenick, M., Tonello, L., Gabrielli, F., Tuszynski, J. A. (2017). Document of Trapani on animal consciousness and quantum brain function: A hypothesis. Journal of Integrative Neuroscience 1–5.

Cocchi, M., Gabrielli, F., Tonello, L., Tuszynski, J. A. (2017). dialogue on the issue of the "quantum brain" between consciousness and unconsciousness. Journal of Integrative Neuroscience 16: S1–S2.

Cocchi, M., Gabrielli, F., Tonello, L. (2015). Molecular and Quantum Approach to Psychopathology and Consciousness. Ann Depress Anxiety 2 (2): 1046.

Cocchi, M., Minuto, C., Tonello, L., Gabrielli, F., Bernroider, G., Tuszynski, J. A., Cappello, F., Rasenick, M. (2017). Linoleic acid: Is this the key that unlocks the quantum brain? Insights linking broken symmetries in molecular biology, mood disorders and personalistic emergentism. BMC Neurosci 18: 38.

Cocchi, M., Tonello, L., Gabrielli, F. (2014). Psychiatric Doubts. Open Journal of Depression 3: 5-8.

Cocchi, M., Minuto, C., Tonello, L., Tuszynski, J. A. (2015). Connection between the Linoleic Acid and Psychopathology: A Symmetry-Breaking Phenomenon in the Brain? Open Journal of Depression 4: 41-52.

Cocchi, M., Tonello, L. (2006). "Biological, Biochemical and Mathematical considerations about the use of an Artificial Neural Network (ANN) for the study of the connection between Platelet Fatty Acids and Major Depression". J. Biol. Res. LXXXI: 82-87.

Cocchi, M., Tonello, L., Gabrielli, F. (2014). Mood Psychopathologies: An Integrated Complexity-Based Interpretation. Psychology 5: 192-203.

Cocchi, M., Tonello, L. (2010). "Running the hypothesis of a bio molecular approach to psychiatric disorder characterization and fatty acids therapeutical choices", Annals of General Psychiatry 9 (supplement 1): S26.

Cocchi, M. (2019). In memory of Dr. Kary Mullis. Journal of Biological Research 92: 8513.

Damasio, A. (2010). Self Comes to Mind: Constructing the Conscious Brain, Pantheon, New York.

Dennett, D. C. (1992). Consciousness Explained. Back Bay Books New York.

Dennett, D. C. (1987). The Intentional Stance. The MIT Press, Cambridge.

Gallese, V., Goldman, A. (1998). Mirror neurons and the simulation theory of mind-reading. Trends and cognitive sciences 12: 493-501.

Gallese, V., Sinigaglia, C. (2010). The bodily self as power of action. Neuropsychologist 48, 3: 746-748.

Heidegger, M. (2010). On the way to Language. Harper Collins Publishers Inc, New York.

Nagel, Th. (2012). Mind and Cosmos: Why the Materialist Neo-Darwinian Conception of Nature is Almost Certainly False. Oxford University Press, Oxford.

Searle, J. R. (1997). The Mystery of Consciousness, New York Review, New York.

Searle, J. R. (1992). The Rediscovery of the Mind. The MIT press, Cambridge.

Summhammer, J., Sulyok, G., Bernroider, G., Cocchi, M. (2020). Forces from Lipids and Ionic Diffusion: Probing lateral membrane effects by an optimized filter region of voltage dependent K+ channels. arXiv. Org. Physics.

Tarlaci, S. (2019). Quantum neurobiological view to mental health problems and biological psychiatry. Journal of Psychopathology 25: 70-84.

Tegmark, M. (2015). Consciousness as a State of Matter.; https://arXiv.org/abs/1401.1219.

Tonello, L., Cocchi, M., Gabrielli, F., Tuszynski, J. A. (2015). On the possible quantum role of serotonin in consciousness. J. Integr. Neurosci. 14: 295.

Tononi, G., Boly, M., Massimini, M., Kock, C. (2016). Integrated information theory: from consciousness to its physical substrate. Nature Reviews Neuroscience 17: 450-461.

Book Review

Review of Wolfgang Baer's Book: Conscious Action Theory - An Introduction to the Event-Oriented World View

James H. Lake[*]

Department of Psychiatry, University of Arizona College of Medicine, USA

Abstract

Wolfgang Baer's project consists of nothing less than adducing from first principles a model of consciousness that reconciles subjective experiences with biophysical processes inherent in brain function (Baer 2020). His model, Conscious Action Theory (CAT) argues that unification of the subjective and objective can only be achieved by accepting the observer's existence as the foundational premise underlying all scientific inquiry. Starting from this premise it becomes possible to abandon object-oriented epistemologies that view subjective experience as epiphenomena of physical processes, and to develop a truly event-oriented ontology in which the first-person observer is a physical system that has always contained its own mental experience. Baer asserts that mind, brain, and world can be unified only when physics evolves to encompass both the causes and *subjective feelings* of experiences. His book transcends mere philosophical speculation by proposing the expansion of contemporary quantum physics to implement these unifying concepts.

Keywords: Conscious action, event, world view, brain, mind, experience, subjective, objective, biophysical process.

Concepts

Baer argues that the Aristotelian paradigm positing an a-priori independent world of dead objects was so strongly ingrained in physics historically that reference to the subjective observer was anathema. Even in the early 1900's Schrödinger's wave function was interpreted as a quantum probability object existing independently of human experience. Subsequently, logical positivism held that physics should be concerned only with relations between observable experimental results, which were statistical in nature, and that efforts to understand underlying causes was misleading and unnecessary. Since those early days, research findings have confirmed that physical effects take place at the quantum level in relationship to the mental experiences of scientists conducting research and observers. These findings have resulted in a paradigm shift from Aristotle's naïve empiricism to worldviews approaching Platonic idealism. The philosopher Whitehead (1959) who had a deep understanding of quantum mechanics, proposed that real events he called actual occasions should comprise the building blocks of reality rather than probabilities as defined by quantum mechanics, or objects as conceived in Aristotelian naturalism. More recently Vitiello (2001) and Tegmark (2014) have argued for the logical

[*]Correspondence author: James H. Lake, MD, Assistant Professor, Department of Psychiatry, University of Arizona College of Medicine, USA. E-mail: jameslakemd@gmail.com

necessity of including the brain and the subjective experiences of the experimenter and observers in any investigation of the quantum nature of reality.

According to Baer the phenomenal content of sensory awareness is only one phase of a larger *processing system* required to generate what is actually *experienced*. This larger self—big "I"—stands in contrast to the phenomenal appearances of our surroundings and the little "i" we actually see and experience as our body. Differentiating between what we see "i-you-it" and what we are "I-You-It" is critical to understanding ourselves as events. The big "I" processing system can be described in strictly physical terms as a cyclic action flow structure. According to the CAT-model, a unity of mind and body exists as a sequence of inner and outer forces contained in event structures. Applying this unity to all objects allows humans to conceive of the rest of the universe as an event which differs from our *human event selves* in magnitude only. We both do the same thing and are in that sense equal partners in the dance of existence.

Baer reminds the reader that philosophers and physicists among them James, Whitehead, Bohm and Penrose have argued for an event-oriented world that places the objective world into the context of an event, which like the big "I" generates feelings of itself. The event-oriented worldview proposes a Universe of interacting action cycles each of which contains its own experiences evolving in its own time-frame. Baer asserts that the next major advance in understanding of the nature of phenomenal reality will be achieved when the universe is conceived as an event in which interacting conscious beings are formally described as action cycles.

Physics

After summarizing important philosophical arguments, the second section of the book develops a sophisticated model for an event-oriented world grounded in the knowledge of contemporary physics. Unity of mind, body and the universe can be achieved within this framework by recognizing that humans are events that contain both mental experiences and associated physical phenomena. Baer postulates that energy fields associated with charge and mass formalized as the weak and strong forces of nuclear physics correlate to observable mental experiences, and that gravito-electric forces correspond to the objective physical world. In this schema, mental and physical phases follow one another manifesting as a web of inter-relationships between subjective experience and external states of affairs *in the world*. This process results in repeating cycles of *mind-causing-body-causing-mind*. The above framework fundamentally changes our view of physics and the objective world it describes. Physics no longer describes 'things out there' but the apparent properties of events experienced by an observer. Everyone generates a unique *world of experience* as a function of stimulation received. Conscious awareness takes place when a measurement function compares signals from an internal model of the external world with computations of a neural network chain processing sensory or cognitive information. Nothing exists for mind and body to interact with outside of these cycles. In sum, the subjective and objective are seamlessly integrated in a universe that has both a physical aspect and a mental aspect.

Implications

The third section of the book is a wide-ranging discussion of implications of Conscious Action Theory for current limited explanatory models of consciousness grounded in neuroscience, artificial intelligence, philosophy, psychology and religion. Baer asserts that equating the action cycle as a continuous happening that takes place in the domain of subjective experience to the physical universe solves the 'hard problem' of consciousness since the physical basis of subjective experiences is, by definition, equivalent to the 'Now' phase of an action loop. Chapter 6 offers a functional description of action flow in a conscious being including many insightful discussions of sensory awareness, memory, and action cycles that result in the generation of meaning. Chapter 7 discusses implications of Conscious Action Theory for neuroscience, artificial intelligence and psychiatry. Baer asserts that the same internal mass-charge structures necessary and sufficient for consciousness to take place are present in all material systems, at least on a primitive level. It follows that there is no inherent limitation of future AI systems that would preclude the emergence of consciousness.

Baer's proposed comparator mechanism implies that the physical correlates of human consciousness are located in the interface between the neuronal and glial networks. Recent research findings (Mitterauer 2020) have identified these interfaces as the field of tri-partite synapses controlled by Astrocyte cells and have confirmed that imbalances between neurotransmitters and receptors underline some psychiatric disorders.. Chapter 8 discusses general implications of Conscious Action Theory for philosophical, psychological and religious understandings of consciousness. Baer concludes the book with a review of ongoing investigations into the nature of consciousness aimed at confirming the existence of consciousness action cycles. Among other innovative approaches, he describes ongoing studies of dual eye and bi-scopic perception, and patented training devices designed to modify our normal three dimensional mental display, aimed at increasing our cognitive information bandwidth to more closely match emerging brain-computer interface technologies (Baer N 2012).

Received June 23, 2020; Accepted July 25, 2020

References

Baer N., Baer W. (2012) "Interest-Attention Feedback System for Separating Cognitive Awareness into Different Left and Right Sensor Displays," Patent Application No 13/455,134, filed April 25, 2012

Baer W. (2020) Conscious Action Theory: an introduction to the event oriented world view, Routledge Press, ISBN: 978-1-138-66746-4 (hbk)

Mitterauer B., Baer W. (2020)"Disorders of Human Consciousness in the Tri-partite Synapses" *Medical Hypothesis* Vol 136, March 2020, URL: https://doi.org/10.1016/j.mehy.2019.109523

Tegmark M., (2014) "Our Mathematical Universe ", Random House LLC , ISBN 978-0-307-59980-3, Chapter 10, p243

Vitiello, G.(2001) *My double unveiled: The dissipative quantum model of the brain*, John Benjamins ISBN 9027251525

Whitehead, Alfred North, (1978) "Process and Reality an Essay in Cosmology", Corrected Edition by D. R. Griffin and D. W. Sherburne, The Free Press N.Y. ISBN 0-02-934580-4